OCCASIONAL
P A P E R

T0302667

# Options for Improving the Military Child Care System

Gail L. Zellman, Susan M. Gates, Michelle Cho,
Rebecca Shaw

Prepared for the Office of the Secretary of Defense

RAND NATIONAL DEFENSE RESEARCH INSTITUTE

The research described in this report was prepared for the Office of the Secretary of Defense (OSD). The research was conducted in the RAND National Defense Research Institute, a federally funded research and development center sponsored by the OSD, the Joint Staff, the Unified Combatant Commands, the Department of the Navy, the Marine Corps, the defense agencies, and the defense Intelligence Community under Contract W74V8H-06-C-0002.

**Library of Congress Cataloging-in-Publication Data**

Options for improving the military child care system / Gail L. Zellman ... [et al.].
    p. cm.
    Includes bibliographical references.
    ISBN 978-0-8330-4414-3 (pbk. : alk. paper)
    1. Children of military personnel—Care—United States. 2. United States. Dept. of Defense—Officials and employees—Salaries, etc. 3. Employer-supported day care—United States. 4. Day care centers—United States. I. Zellman, Gail.

UB403.O72 2008
362.71'2068—dc22

                                                                    2008010143

The RAND Corporation is a nonprofit research organization providing objective analysis and effective solutions that address the challenges facing the public and private sectors around the world. RAND's publications do not necessarily reflect the opinions of its research clients and sponsors.

**RAND®** is a registered trademark.

Published 2008 by the RAND Corporation
1776 Main Street, P.O. Box 2138, Santa Monica, CA 90407-2138
1200 South Hayes Street, Arlington, VA 22202-5050
4570 Fifth Avenue, Suite 600, Pittsburgh, PA 15213-2665
RAND URL: http://www.rand.org/
To order RAND documents or to obtain additional information, contact
Distribution Services: Telephone: (310) 451-7002;
Fax: (310) 451-6915; Email: order@rand.org

# Preface

The U.S. military child care system provides one of the largest in-kind benefits that the U.S. Department of Defense (DoD) offers. DoD supports the largest employer-sponsored system of high-quality child care in the country. A fraction of the military members who use the system receive a large in-kind subsidy. As a system of employer-sponsored care, however, DoD child care ultimately must be assessed in terms of its contribution to military goals—in particular, recruiting, readiness, and retention.

The Office of the Under Secretary of Defense for Personnel and Readiness, which produces the Quadrennial Review of Military Compensation, asked RAND to assist in its efforts to improve the effectiveness of the compensation and benefits system for the uniformed services by examining the current military child care system as a component of that larger system. This paper reviews system costs and outcomes, examines private-sector employer responses to employee child care needs as a source of comparison, and considers a number of potential changes to the military system. These changes hold promise for better meeting the goals of DoD as an employer, providing military parents with more choices, distributing benefits more widely, and improving the quality of care for children. This information will help the Office of the Under Secretary of Defense for Personnel and Readiness to understand the issues and challenges inherent in a new paradigm in which military child care is considered a compensation issue, and the potential of this paradigm is assessed from a broader, more employer-based perspective.

This paper will be of interest to officials responsible for military compensation, military readiness, and military retention and to those concerned about the availability and quality of military child care.

This research was sponsored by the Office of the Under Secretary of Defense for Personnel and Readiness and conducted within the Forces and Resources Policy Center of the RAND National Defense Research Institute, a federally funded research and development center sponsored by the Office of the Secretary of Defense, the Joint Staff, the Unified Combatant Commands, the Department of the Navy, the Marine Corps, the defense agencies and the defense Intelligence Community.

For more information on RAND's Forces and Resources Policy Center, contact the Director, James Hosek. He can be reached by email at James_Hosek@rand.org; by phone at 310-393-0411, extension 7183; or by mail at the RAND Corporation, 1776 Main Street, Santa Monica, California 90407-2138. More information about RAND is available at www.rand.org.

# Contents

# Figures

# Tables

# Summary

The U.S. military child care system is the largest employer-sponsored child care system in the nation, widely recognized for providing high-quality care. A range of different settings enables the system to meet military parents' needs for reliable, high-quality care while recognizing parental preferences concerning environment, size (the number of children cared for in that provider setting), and flexibility. Subsidies based on family income ensure affordability.

## Study Motivation

Despite its size, the military child care system serves only a small percentage of eligible families needing child care assistance. Care in Child Development Centers (CDCs) is quite costly for DoD to provide; care for the youngest children is particularly expensive since parent fees are based on family income and not on the cost of care. Care in Family Child Care (FCC) homes is substantially less costly. There is little evidence that the care provided in DoD-run CDCs and FCC homes addresses DoD employer goals of increased readiness, retention, and recruitment. Moreover, families that cannot or choose not to use CDC or FCC care receive no help covering their child care expenses. Moreover, they may rely on care that is mediocre, given their often limited financial resources and the fact that the average quality of care in civilian communities is generally not high. The Office of the Under Secretary of Defense for Personnel and Readiness asked RAND researchers to use the insight they have gained during several previous studies on military child care (e.g., Zellman and Gates, 2002; Moini, Zellman, and Gates, n.d. and 2006) to reexamine military child care as a compensation issue and evaluate options for transforming the current military child care system.

## Methods

In this paper, we provide an overview of the military child care system and assess the system's success in cost-effectively meeting DoD readiness, retention, and recruitment goals. In particular, we consider the logic of DoD offering military child care as an in-kind benefit. This assessment is based primarily on a review of existing research. We supplemented our own extensive prior research with a review of private-sector employee child care benefits and interviews with a small number of companies that are similar to DoD in important ways.

## Findings

### Child Care in the Private Sector

Private employers provide child care benefits with an eye to the bottom line: They offer these benefits to improve recruitment, reduce absenteeism, and decrease turnover. Some employers operate child care centers or subsidize care in the community; many provide resource and referral services. In recent years, employers have begun to offer benefits of a different kind—moving away from specified services and goods that the employer pays for, such as child care centers, to changes in the work environment that employees value highly, such as flextime and the ability to work from home. However, these flexible benefits do not necessarily obviate the need for child care; further, they fail to address two chronic problems in locating and using care: lack of availability and mediocre quality.

### Child Care in the Military

Military child care programs are reaching a small fraction of the total military population. At most, 7 percent of military members are served by CDCs, and another 4 percent by FCC homes. Even among families with children under age six, fewer than half use DoD-sponsored child care. Nevertheless, the vast majority of the child care resources spent by DoD are devoted to care provided in CDCs. In general, the cost of CDC care is substantially higher than the cost of FCC or the cost to DoD and military parents (via co-payments) for child care provided by civilian contractors. Moreover, CDC care, with high fixed hourly costs, is inherently less flexible than FCC. In homes, FCC providers theoretically can accommodate duty hours that exceed those of the CDCs and provide children with more continuity of care.

Our evidence indicates that child care is a readiness and retention issue for many service personnel. Military members report that child care issues prevent them from reporting to duty and cause them to be late for or absent from work. Some military members also indicate that child care issues may lead them to leave the military.

Despite frequent assertions by DoD that the key goals of the military child care system are the promotion of readiness and retention, the system is not organized to effectively promote these goals or to monitor the extent to which they are being addressed. CDC enrollment priorities for certain family types—e.g., single parent and dual military—are designed to promote these goals, but the evidence suggests that they are not working effectively. The Military Child Care Act of 1989, passed in response to child abuse allegations in military child care centers, focuses the system (and its resources) on protecting children in DoD CDCs, providing them with high-quality care, and increasing the availability of care through large subsidies to CDCs. Even if CDC care were enhancing readiness and retention among those families that use it, the overall effect of the CDCs on these objectives would be limited because so few military families actually use the CDCs. However, there are reasons to doubt that CDCs are having any positive effect, even for those families that use them. Of particular concern is that when surveyed and when we controlled for family type, families that use CDCs were actually more likely than other families to report that they were likely to leave the military because of child care issues. Moreover, many of those who receive the large CDC subsidy are unaware of its value; some even believe DoD is profiting from their CDC fees. These factors limit the extent to which the subsidy can promote retention.

The evidence presented in this paper raises concerns that the current system is not meeting DoD or service member needs in an optimal way. DoD appears to be reaping limited benefits from the large CDC subsidy, while many military parents get no help at all.

## Recommendations

Our findings suggest that the DoD child care system could change in a number of ways to better meet DoD and military family needs. For example, it could redistribute resources within the current system. In pursuing this strategy, DoD potentially would be able to provide military benefits to more families and provide the types of care that would be more likely to improve readiness. A redistribution of resources could involve redirecting money from CDCs to FCC, targeting the child care benefit to different types of families, or focusing the benefit on different types of care, such as care in local communities.

Rethinking priority policies from the perspective of *both* child care need and the degree to which care characteristics fit with likely DoD and service member needs would be another important way to change the system. DoD may also wish to redistribute resources in order to allocate child care benefits to more families in more settings. To do this without increasing overall expenditures would involve a reallocation from those who are currently receiving a large subsidy to those who are receiving little or no subsidy. Alternatively, DoD may wish to redistribute benefits by targeting them more narrowly to those families that value them the most, such as military members who are deployed, or to the families of those it values most highly and whom DoD is most concerned about losing, such as individuals with special skills.

In support of recruitment, readiness, and retention goals, DoD may also wish to expand the child care benefit to cover more military families and a broader set of child care needs. DoD could consider expanding DoD-provided care and evaluate the system in terms of availability of care and contribution to readiness. Such an effort would likely be costly, although costs could be moderated by expanding FCC and moving the care of the youngest children out of CDCs, focusing center-based care on older children. However, it is unlikely that expanded CDC care or FCC would meet the needs of all military families, many of which live far from the installation.

Alternatively, DoD could expand access to child care through the use of cash benefits, vouchers, and/or negotiated discounts with local providers, while continuing to provide some amount of FCC and CDC care. In the long run, this approach likely would also increase child care availability and the average quality of care that DoD dependents receive. If DoD chose to focus on improving local provider quality, it could exert a substantial positive influence on the overall quality of care in communities surrounding military installations. These efforts might enable more military parents to use higher-quality care, which could improve readiness and retention.

DoD may want to invest more resources in assessing the value of child care benefits, as it does for other military compensation components. To prepare for such analyses, DoD should track and centrally record information (e.g., rank, specialty, family type) about parents who enroll their children in the child care system and the amount and type of care being used. This information could then be used to assess the effect of system utilization on readiness and retention. Ideally, DoD would also retain information on the children being served in order to facilitate assessments of near-term child outcomes (e.g., health status, test scores from first grade).

DoD might also want to consider conducting a periodic child care needs assessment to understand changing family needs and preferences. It would also be important for DoD to regularly assess the private-sector child care market in local military communities. These data would inform the key decision of whether to promote the development and use of high-quality care off base.

## Conclusions

The DoD child care system provides high-quality care to a small percentage of military members, with seemingly limited payoff in terms of readiness, retention, and recruitment. DoD can do a better job of addressing the child care needs of military families and its own needs for a stable, ready force by rethinking the current system, collecting important data on utilization, and examining the link between utilization and key employer outcomes.

# Acknowledgments

The authors would like to thank Denny Eakle and Saul Pleeter for supporting this work and drawing attention to military child care as a compensation issue. For information on private-sector trends and motivation for providing child care benefits, we gratefully acknowledge the important contributions from Nancy Costikyan of Harvard University, Tammy Palazzo of *Working Mother,* Betty Purkey of Texas Instruments, Kathy Truscott of Daimler Chrysler Corporation, and Ed Willis of Union Pacific Railroad. We benefited from numerous conversations with them. We also thank participants at the Western Economic Association conference in Seattle, Washington, for many helpful comments. We are grateful to Rebecca Kilburn of RAND and Carla Tighe Murray of the Congressional Budget Office for helpful reviews of a prior draft of this paper. We also thank Michael Kuzara of the U.S. Military Academy, Neil Singer and Greg Davis from IDA, and Sydne Newberry of RAND for their comments on an earlier draft. Christopher Dirks assisted with document formatting and editing. Any errors are the sole responsibility of the authors.

# Abbreviations

| | |
|---|---|
| BLS | Bureau of Labor Statistics |
| BRAC | 2005 Base Realignment and Closure |
| CDC | Child Development Center |
| DCRA | dependent care reimbursement account |
| DoD | U.S. Department of Defense |
| EAP | employee assistance program |
| FCC | Family Child Care |
| FEEA | Federal Employee Education & Assistance Fund |
| GAO | Government Accountability Office |
| GSA | U.S. General Services Administration |
| MCCA | Military Child Care Act of 1989 |
| NACCRRA | National Association of Child Care Resource and Referral Agencies |
| NAEYC | National Association for the Education of Young Children |
| OPM | U.S. Office of Personnel Management |
| QRS | quality rating system |
| R&R | resource and referral |
| SAC | school-age care program |

# CHAPTER ONE

# Introduction

## Background

The U.S. military child care system is the largest employer-sponsored child care system in the nation, and it is widely recognized for providing high-quality care. Its current form and scope belie its humble origins in informal and locally based efforts to provide child care for a few hours here and there while military wives undertook volunteer or personal activities. The system began to change and grow as more and more military spouses sought employment outside the home and as single-parent and dual-military families became more common in the military. However, this growth was largely unstructured and unregulated, so that the quality of care provided varied considerably across installations.

Allegations of child abuse in military child care programs focused congressional attention on the quality of the care being provided by the U.S. Department of Defense (DoD). The Military Child Care Act of 1989 (MCCA) formalized the system through a set of regulations to ensure quality and increase availability through substantial, targeted subsidies. These regulations concentrated the system's attention on providing high-quality child care in Child Development Centers (CDCs) operated by DoD.

Today, military child care is provided as part of a system of care designed to meet the needs of military families as children age. Children served by the system range from six weeks to 12 years old. A variety of different settings enables the system to meet parents' needs for reliable, high-quality care while recognizing parental preferences concerning environment, size (number of children cared for), and flexibility. The military provides care for as many as 12 hours a day in CDCs and even longer, if necessary, in Family Child Care (FCC) homes. For those families with more limited needs, care may also be provided on a part-time and an hourly basis in CDCs and FCC homes in many locations. Before- and after-school programs are also available to care for school-aged children in a center-like setting; youth programs provide a relatively unstructured but supervised setting for older school-aged children. Consistent with MCCA, there have been substantial efforts to increase affordability and availability of child care. The military child care system, while certainly a benefit to those who use it, has never been viewed as an entitlement (i.e., a benefit that is provided to all who meet predetermined eligibility criteria) in the way that health benefits or housing benefits have been. Nor has it been focused on addressing key employer goals: enhanced recruiting, retention, and readiness, despite the considerable subsidy the military provides.

## Motivation for This Study

Previous RAND studies have revealed that the heavily subsidized care provided in DoD-run CDCs and the care provided in FCC homes benefit only a small percentage of eligible families needing child care assistance. Families that cannot or choose not to use CDC care or FCC receive no help in paying for child care. In addition, many families using CDC care do not recognize the value of the substantial subsidy that they are receiving. Those studies have questioned whether the current military child care system is meeting the needs of DoD and military families in the most effective and equitable manner.

The Office of the Under Secretary of Defense for Personnel and Readiness asked RAND to use its knowledge about military child care to assess the military child care system from a compensation perspective and to consider ways in which the system might better address DoD recruitment, readiness, and retention goals.

## Context for Considering Military Child Care as a Compensation Issue

To assess the military child care system from a compensation perspective, it is important to consider three distinct questions. First, what are the goals of the military compensation system and how does military compensation differ from compensation in the civilian sector? Second, why do employers ever provide noncash, in-kind, or deferred cash benefits in lieu of cash compensation? Third, given the answers to the first two questions, what is the logic for providing military child care support as an in-kind benefit?

### The Military Compensation System

With the transition from the draft to an all-volunteer force in the early 1970s, compensation became an important tool through which DoD manages recruitment, readiness, and retention (Rostker, 2006; CBO, 2007). In an all-volunteer force, individuals must be convinced to join and remain in the military and to separate when it is in the best interests of the organization; the compensation system is a key lever to influence those decisions (Hogan, 2004). The impact of compensation on the decisions of military personnel is moderated by a number of factors outside the military, particularly the strength of the U.S. economy, employment opportunities in the private sector, and the changing demographics of new recruits. In recognition of these factors, in the early 1980s, DoD began to adjust the levels of military compensation to reduce the pay gap between military and civilian compensation (CBO, 2007). For example, in the late 1990s, when a booming economy and post–Cold War downsizing policies contributed to large drops in military recruitment and missed recruitment targets, the military increased cash compensation levels and expanded retirement benefits (Williams, 2004; CBO, 2007). These changes have been credited for at least part of the increased recruiting levels between 1998 and 2003 (Williams, 2004). Studies have also found that compensation plays a role in military members' decisions to stay in the military, rather than retire or separate early (GAO, 2000).

From the inception of the all-volunteer force, the military compensation system has been subject to assessment and evaluation, most notably through the Quadrennial Reviews of Military Compensation. The purpose of these reviews is to ensure that DoD is making the most effective use of its compensation tools (basic pay, benefits, bonuses, and recruiting incentives) in the most cost-effective way to meet its recruiting, retention, and readiness goals. These reviews are particularly important given that compensation costs account for approximately 30 percent of DoD's budget (Williams, 2004).

The military compensation package is made up of the following components: basic pay, special and incentive pays, retired pay and other deferred benefits, housing, food provided in-kind or through cash allowance, health care, child care, and an array of other in-kind benefits that fall under the category of morale, welfare, and recreation (Williams, 2004; CBO, 2007). Basic pay rates are based on rank and length of service, with limited variation based on performance or other factors. The primary rationale behind limiting variation in cash pay is the promotion of organizational solidarity; it is believed that substantial variation could lead to lack of order and discipline (Williams, 2004). Other pay does vary based on family status, location of service, and types of duties being performed (e.g., hazard pay, submarine pay, and similar special pays). For example, military personnel with dependents receive a basic allowance for housing that has been reported to be between 11 and 30 percent higher than those without dependents (Hogan, 2004; CBO, 2007). A range of other bonuses and special compensation exists to encourage recruitment and retention, though they include only about 4 percent of total cash pay (Hogan, 2004). This pattern contrasts with that in the private sector, where individual bonuses for professional and middle-management personnel can be in the range of 10–30 percent of an employee's compensation and, unlike in the military, are more closely and easily tied to specific performance measures (Strawn, 2004). Inflexibility in the compensation system is the most common criticism of the military compensation system and is the major difference between the military and private sector (Williams, 2004; Hogan, 2004).

Traditionally, military members have received a much larger share of their compensation through in-kind benefits—such as housing, meals, and services—than through direct cash compensation, compared with their civilian counterparts. That has been changing gradually over the years as the military has moved toward providing more benefits in the form of cash allowances, such as housing allowances, that had formerly been provided in-kind. Still, according to a recent report by the Congressional Budget Office, findings of which have been supported by other studies, the typical enlisted person receives approximately half of his or her total compensation in noncash and deferred cash benefits, compared to 33 percent for civilians (CBO, 2007; Strawn, 2004). The amount of compensation provided through noncash benefits has increased in recent years. As in the private sector, subsidized health insurance makes up the largest portion of employees' noncash benefits (Murray, 2004; CBO, 2007).

Another important difference between military and civilian compensation is that the military system places a greater emphasis on equity and fairness, with basic pay being driven primarily by longevity or years of service; performance plays a limited role. Supplements to basic pay are provided to compensate individuals for particular types of duty, e.g., hazard pay, and are offered to everyone who works under these conditions. Two important exceptions are made to the general rule of equity. One concerns special compensation provided to individuals with special skills, such as pilots or physicians. This compensation is designed to retain these highly trained individuals, who are believed to have more options in the civilian sector than other military members. A second exception concerns the treatment of members with families. In spite of its focus on equity, the military, by providing higher base pay and larger housing benefits to members with dependents, compensates members with dependents at higher levels than it does their single colleagues. One rationale that has been offered for this practice is that providing such noncash benefits promotes military readiness by reducing some of the strains associated with family life, such as having to identify easily accessible, affordable, high-quality child care services. This issue is particularly a concern for families that have relocated to a new community as a result of military service. This distinction based on family status has been the subject of much debate on equity grounds.

The move toward providing a larger share of benefits in the form of cash allowances has in many ways made the compensation system more efficient; receiving cash instead of non-cash benefits enables military members to spend their money on what they value most (CBO, 2007). However, receiving cash raises the likelihood of dissatisfaction, because military members who previously received a highly valued noncash benefit, such as on-base housing or child care, discover that the cash benefit does not allow them to purchase services of equal quality on the open market.[1]

Researchers have offered a range of recommendations for ways to change the military compensation system to enable it to better meet the military's goals for recruitment, productivity, and retention of valuable personnel. Examples include varying pay by military occupation rather than rank in order to reward personnel based on more highly valued and/or technical skills, increasing pay based on the time spent in a grade rather than the overall time in service (which would reward high achievers who were promoted ahead of schedule), making more efficient use of special pays and bonuses to meet recruitment and retention targets for specific personnel, and improving the structure and delivery of in-kind benefits to military members. All of these changes would result in a more flexible compensation system that likely would improve the military's ability to attract valuable recruits (Hogan, 2004; Murray, 2004).

## Why Do Employers Offer In-Kind Benefits?

As mentioned above, a much larger share of military compensation comes in the form of in-kind, targeted or deferred-cash benefits, compared with civilian compensation. It is worth considering why employers provide benefits. Economic theory suggests that in general, cash is the most cost-effective form of compensation because the employee is then able to allocate that cash to the goods or services that provide him or her with the greatest marginal return. By providing in-kind compensation or otherwise restricting the way in which compensation can be provided, there is a risk that one dollar spent on a given benefit has a lower value than one dollar received by the employee in cash.

There are a number of conditions under which it is economically efficient for employers, whether in the military or private sector, to provide in-kind benefits. These conditions include situations in which employers are able to provide the benefit at a lower cost than if the employee were to obtain the goods or service individually (e.g., group health care); the benefit allows the employer to screen for or encourage certain characteristics in their employees (e.g., fitness level); the composition of the compensation package matters to the employer (e.g., child care provided to employees' dependents is of high value because high-quality care more effectively promotes readiness); the benefit engenders good will and loyalty to the employer; and the benefit is seen by employees as likely to be more stable (i.e., not likely to be taken away), as in the case of commissaries (Murray, 2004; CBO, 2007). In the case of the military and

---

[1]    On-base housing is a useful example of this situation. Military personnel have the option of choosing either a housing allowance or on-base housing, the latter of which is subject to availability. A 1999 RAND study revealed that military members viewed the in-kind benefit as more economically valuable than the cash allowance. That study also revealed that those who used on-base housing preferred it primarily because of its economic value and not because of other factors, such as promoting military values or building a sense of community (Buddin et al., 1999).

the private sector, there is some evidence that in-kind benefits and other incentives influence recruitment/hiring and retention decisions of military members/employees, though much of the evidence is based on survey data in which reported intentions are not always matched with actual behavior (Hansen, Wenger, and Hattiangadi, 2002; Buddin, 1998; Golfin, 2003).

Of the reasons listed above, there are two that stand out. The first and most common reason why employers provide in-kind benefits is that the employer can sometimes provide the employee with a value that exceeds the employer's cost of providing the benefit. The U.S. tax code creates many opportunities for such leverage. Certain benefits, such as health insurance or dependent care assistance, are tax free. Similarly, an employee who receives meals at work or whose children attend a child care center that is subsidized by the employer does not pay taxes on the value of those meals or the amount of that subsidy (Murray, 2004). Employers might also be able to provide employees with greater value if they simply happen to be more efficient providers of a particular good or service or if they can take advantage of economies of scale. For example, it is well known that airlines provide their employees with free air travel on a standby basis. Such a benefit is relatively costless for the airline to provide and valuable to the employee.

The military compensation system has been criticized for its heavy reliance on in-kind benefits. Critics charge that it costs the government more to provide the good or service directly than it would to provide a cash allowance to the individual to obtain the good or service on his or her own. Related to this, costs are likely to rise as base infrastructure ages. Another criticism is the inequity that accompanies those in-kind military benefits that are focused on base and, therefore, are not accessible to the many military personnel who do not live on or near an installation (Williams, 2004). It is also argued that the differential pay that the military provides based on family status creates an incentive for military personnel to marry and have children, in spite of the strain that military service can place on a family. Indeed, there is evidence that the proportion of married people, by age, is higher in the military, compared with the private sector. While providing important support to military families, these additional benefits may actually weaken readiness capabilities and encourage less desirable military members to stay in the military (Hogan, 2004; Raezer, 2004).

The other main reason why employers provide employees with in-kind benefits is because they want to influence the choices that employees make. Often, employers want to do this because the choices have ramifications for employee productivity. For example, employers may want all employees to have health insurance whether employees want it or not because the employer believes that when employees have health insurance, they are healthier and more productive. An employer who feels strongly about the productivity benefits of health insurance may, therefore, provide it and not allow employees to opt out of coverage even in the absence of tax considerations.

### Military Child Care as an In-Kind Benefit

To assess military child care as a compensation issue, it is necessary to understand why DoD chose to provide child care as an in-kind benefit. An important reason for this choice may be found in the widespread understanding that child care availability and quality can affect key DoD goals of readiness, retention, and possibly recruitment. Some military members with young children need or want child care for their children. For some of these families, access to child care may affect the ability of the military member to show up for duty. This need may be particularly true for dual-military and single-parent families, which lack a parent who can be counted on to provide backup care. Moreover, military families may need care at times when civilian care is difficult to find or simply not available. Military members are often required to report for duty on nights and weekends. They may be called to duty on short notice; they may have irregular shifts or shifts that do not correspond with the typical workday. Concerns about the quality of child care may affect a parent's ability to concentrate on their job. If quality is assured, the parent can more comfortably focus on the work at hand.

Deployments pose an additional set of issues for military families with young children. Spouses who work outside the home may need additional child care support to manage life as a working parent while the military member is deployed. Spouses who do not work outside the home may need access to child care at various times of the day in order to get household chores done or simply to have some respite. The needs of families affected by deployment may be quite different from the needs of other military families.

Ultimately, child care can become an important retention issue—particularly at a time of active deployment. If families become frustrated with military life because of a lack of child care options that meet their needs, they may decide to leave the military entirely. Child care support may also factor into the decision of recruits—particularly older recruits who already have families or may be thinking about having families.

The above discussion suggests reasons why DoD, as an employer, might need or want to provide additional support to members with young dependents in the interests of recruitment, readiness, and retention. Furthermore, it suggests reasons why DoD might want to provide reliable high-quality child care that is flexible enough to meet the demands of military life rather than simply giving families additional cash to spend as they wish. Indeed, DoD does already provide higher pay to military members with dependents, as discussed above. By offering an additional benefit that directs families with young children to spend a certain amount of money on a limited set of child care providers who meet military quality standards, DoD may be able to ensure that the child care that families use is reliable and flexible (hence contributing to readiness), that the care is of high quality (hence contributing to readiness, retention, and possibly recruitment), and that spouses of deployed military members actually do get the additional support they need during times of deployment (hence contributing to retention).

There are several reasons why DoD might decide to provide child care directly, rather than provide an allowance or a voucher that could be used to purchase child care services. One reason would be that DoD can provide child care more efficiently than other providers. Another argument is that because of shift work and irregular schedules, military families have unique child care needs that cannot be easily met by private-sector providers. A related argument is that there are not enough high-quality providers in the private sector to serve all military families. In assessing arguments about a lack of private-sector care with particular characteristics, it is necessary to consider not only what the market for child care currently looks like, but also what it could look like if DoD were to provide child care support through other means.

Any child care benefit raises some important fairness and distribution issues for DoD to consider. Should such a benefit be available to all military members with young children, or is it acceptable for DoD to offer benefits to some military families but not others? If inequitable support is offered, what is an appropriate basis for allocating the benefit and varying its magnitude?

The system that has developed in response to MCCA has not considered the goals embedded in that legislation. Certainly, the mandates of MCCA and its focus on centers has played a major role in the continuing focus on CDCs in terms of regulation, subsidies, and policy. In recent years, DoD has begun to support other child care programs, but they do not receive the same financial support or attention that is devoted to the centers. For example, the Navy has developed and tested the 24/7 Group Homes program, which is able to provide round-the-clock care to children of all ages. The Air Force Home Community Care program provides free in-home quality child care services to Air National Guard and Air Force Reserve members during scheduled drill weekends in local communities. The Army has worked to create community-based program options for school-aged children, including access to Boys & Girls Clubs and 4-H Clubs in all 50 states. The Air Force has also formed a partnership with the National Association of Child Care Resource and Referral Agencies (NACCRRA) to expand the supply of FCC slots in local communities. These community-based programs formed by local child care resource and referral (R&R) agencies replicate the Air Force's FCC program, thus ensuring quality. NACCRRA (2007) reports that, at the time of the final program evaluation, the Quality Family Child Care program had created an additional 1,250 high-quality child care spaces. The system attempts to address some of DoD's goals as an employer, i.e., promoting readiness and retention. But most efforts are limited to managing the priority for CDC spaces—which, given their high level of subsidy, are widely sought after—or offering child care during installation-wide exercises. Nor does policy or programming address other potential system goals, such as equity in child care benefits or ensuring the quality of care received by dependents using nonmilitary child care.

## Methods

To better understand how private-sector employers think about employee benefits and how they decide which child care and other benefits to provide, we conducted a literature review and interviews with a small number of employers.

### Literature Review

We searched a variety of literatures to find out what benefits are being offered to employees and what motivated employers to offer those benefits. We first searched academic literature—mostly business and economic journals—for employers' perspectives on offering child care benefits. We also conducted Internet searches using terms such as employee benefits, child care benefits, and business case. We reviewed the publication lists of major human resources and benefits consulting companies and several nonprofit organizations focused on supporting families. We identified additional relevant references by scanning the bibliographies of major reports. We also reviewed the literature on the distribution and value of child care benefits. Although we searched the academic literature, most of the information available on private-sector benefits and employer motivation for providing these benefits was found in the literature from benefits consulting companies and public policy organizations working to promote family-friendly policies. Other reviews of similar information have relied heavily on surveys of large, private-sector companies conducted by nationally recognized consulting and research firms. These surveys typically describe what large, private-sector companies are offering in terms of pay and benefits, but some of the surveys may use nonrepresentative samples and less rigorous analytic methods than are typically found in the academic literature. Altogether, we reviewed a total of 52 articles and 30 Web sites.

**Qualitative Interviews**

We identified a group of employers that are known to support family-friendly policies and programs. To find these employers, we first interviewed the staff at Working Mother Media, who annually compile and publish the 100 best companies for working mothers in the magazine *Working Mother*. This interview provided us with an understanding of how the list of 100 best companies is developed. Using the list, we identified several companies to interview based on their similarity to DoD in terms of their workforce and work demands, the child care needs of their employees, and the scope of benefits they provide. We developed a semistructured interview protocol that focused on the employers' motivation for and experience with offering work-life benefits, and particularly child care benefits. Of the dozen companies we targeted, we succeeded in interviewing half of them.

**Analysis**

Interview notes and study findings were systematically reviewed for consistency, differences by industry type, and emerging trends. We used this analysis as a basis for the summary of the ways in which family-friendly companies approach the issue of employer-sponsored child care and how that approach has changed over time. We also assessed the similarities and differences between the way in which private companies and DoD approach employer-sponsored child care.

We used this review along with our extensive prior research on the military child care system to evaluate whether a targeted child care benefit makes sense for DoD and, if so, in what form.[2] We also considered what additional information or system changes might be needed to ensure that a child care benefit is structured effectively.

We address four questions: How do civilian employers approach the issue of child care benefits for employees and what lessons can be drawn from that experience? What is the nature of the current military child care benefit and what goals does it serve? To what extent does the current system promote the goals of recruitment, readiness, retention, and equity? How might child care resources be more effectively directed to support DoD goals?

## Overview of This Paper

The next chapter describes child care as a compensation issue. It reviews evidence on how and why private- and public-sector employers provide child care benefits. The third chapter provides an overview of the military child care system, including what we know about who uses the system and which options they use. The fourth chapter considers the military child care system from the DoD perspective, assessing the cost of different child care benefits to DoD and evidence on the extent to which child care assistance contributes to readiness and retention. The final chapter considers options for transforming the military child care system and offers conclusions and recommendations.

---

[2]  It is important to note that our research preceded the advent of some of the newest subsidy programs that are described in this paper. Hence, our cost and survey data do not reflect these recent efforts by the services to support such community-based child care for military dependents as Military Child Care in Your Neighborhood.

# CHAPTER TWO

# Child Care Benefits in the Private and Public Sectors

Employers in both the private and public sectors recognize that working parents have child care needs and that employers are affected by these needs in a variety of ways. It is widely believed that employers who address these needs are better able to attract high-quality workers, encourage high productivity, elicit job satisfaction in current employees, and retain high-quality workers. A growing number of lawsuits against employers for violation of the Family and Medical Leave Act[1] when workers seek time off or other support to meet family responsibilities may also be sensitizing employers to the need to pay attention to employees' efforts to balance work and family responsibilities. A recent *New York Times Magazine* article (Press, 2007) notes that the "flood" of such lawsuits reflects not only the increased numbers of women in the workplace but the increasing challenges many Americans at all levels of the employment hierarchy seem to be experiencing as they attempt to balance work and family demands.

Through conversations with private- and public-sector employers and a review of the literature on trends in employee benefits and motivations for offering such benefits, we sought to identify employer motives for supplying child care and other benefits and to clarify the importance of child care benefits relative to other work-life benefits. This effort, we believed, would help to identify different ways of thinking about child care benefits and to contextualize our recommendations for new ways of thinking about military child care.

## Employer Motivation for Providing Child Care Benefits

As increasing numbers of women began to enter the workforce in recent decades, child care became a growing concern for employees and employers alike. Recognizing this need through the provision of child care benefits makes an employer more attractive to prospective employees.

The employers with which we spoke confirmed that their primary motivation for providing child care benefits and other work-life benefits is to attract and retain high-quality employees. Indeed, the *2005 National Study of Employers* conducted by the Families and Work Institute found that almost half (47 percent) of employers reported that they implemented work-life policies, including policies related to child care, to recruit and retain employees. One quarter (25 percent) of employers responding to this study reported that they implemented such policies in order to enhance worker productivity and commitment (Bond et al., 2006). Many uni-

---

[1] The Family and Medical Leave Act of 1993 (Public Law 103-3) requires employers with 50 or more employees to allow employees 12 weeks of unpaid leave for specific reasons, such as the birth or adoption of a child or to care for a seriously ill immediate family member.

versities, for example, use child care benefits, such as reserved spots in on-campus child-care centers, as one way to compete for high-caliber candidates. A guaranteed slot in a high-quality child care center on campus can be the factor that swings the candidate's decision from one university to another.

Generally speaking, companies provide child care benefits and other family-friendly policies because they believe that these programs address real employee needs that have ramifications for the employer as well. The *2005 National Study of Employers* found, for example, that 39 percent of employers reported that they implemented family-friendly policies to support employees and families. Various compensation consulting groups and advocacy groups have conducted analyses that suggest an association between work-life benefits and employee satisfaction and turnover. These studies also highlight the desire of private employers to measure the effects of benefit policies and make a "business case" for the benefits they offer. The *Rewards of Work Study*, conducted every three years since 1997 by the Segal Group, sheds light on what "attracts, motivates, and retains" U.S. private-sector employees. Survey participants are randomly selected U.S. workers from the public and private sector who have workplace email access. The latest survey, which included 1,238 participants, shows that a relatively high proportion of employees are satisfied with their benefits (Segal/Sibson, 2006a, b, and c). The study found a positive relationship between satisfaction with benefits and employee engagement (defined by Segal/Sibson as "knowing what to do at work [vision] and wanting to do the work [commitment]"). Data from the Bright Horizons investment impact study looked at several organizations that sponsored child care centers and found that employees who used the centers had a 50 percent lower voluntary turnover rate, compared with those who did not use the centers (3.7 percent versus 7.2 percent), which translated into a cost savings of $3.4 million for employers (Bright Horizons, 2004). A 2002 Families and Work Institute study found support for providing work-life benefits. By looking across the different study time periods, it was observed that as employees increasingly had access to work-life benefits, such as flexible scheduling, they also had higher job satisfaction, greater commitment to their employer, and higher job retention. Employees also reported that they were less often negatively affected at work by circumstances in their personal lives (Bond et al., 2002).

In making a business case for work-life policies, many companies are analyzing return on investment, or the link between spending on work-life policies and business profitability. Such return-on-investment studies provide evidence that there can be sizable returns from work-life policies, including those that support child care needs. A striking example comes from a Bright Horizons (2004) study that compared the performance of Standard & Poor's 500 companies with "family supporting companies" and found that the supportive companies outperformed the others for the three years that they were observed. Similarly, a Watson Wyatt Worldwide survey (n.d.) gathered information from 405 NASDAQ (National Association of Securities Dealers Automated Quotations) and NYSE (New York Stock Exchange) companies about their human resource practices, employees' attitudes about their jobs and workplaces, and financial data. The study found that companies that scored high on a human capital index and whose employees had high levels of commitment had higher shareholder values.

The case for addressing child care–related issues, in particular, is based on the contribution of child care issues to absenteeism and productivity concerns. According to *The 1997 National Study of the Changing Workforce,* 29 percent of employed parents experienced a child care breakdown in the past three months; those child care breakdowns resulted in absenteeism, tardiness, and reduced concentration at work (Bond, Galinsky, and Swanbert, 1998). To address such situations, many companies offer backup child care assistance. Morgan Stanley, for instance, offers 80 free in-home hours of backup care and space at 15 national backup child care facilities ("2006 100 Best Companies," n.d.).

## Employer Provision of Child Care Benefits

Data from the Families and Work Institute surveys show steady support for providing certain kinds of child care assistance among private-sector employers. The most common types of child care assistance provided were a dependent care assistance plan (45 percent of companies) and R&R services (34 percent). Additional types of assistance included on- or near-site child care (7 percent), backup care (6 percent), sick care (5 percent), and voucher or subsidy programs (3 percent). According to Bond et al., 2006, no statistically significant differences were found in the prevalence of child care assistance offered in 1998 compared to 2005.

The Bureau of Labor Statistics (BLS) provides data on the prevalence of employer-sponsored child care benefits offered by private companies. Table 2.1 displays the prevalence of child care benefits among private-sector employees. In 2007, 15 percent of private-sector workers had access to some type of employer assistance for child care. However, more indirect (and cheaper) R&R support is much more common (11 percent) than employer-provided child care subsidies (3 percent) and employer-provided child care (5 percent). While access to child care benefits appears to have increased since 1999 (from 6 to 15 percent of employees), this increase is largely the result of the recent inclusion of additional options, such as R&R.

Other surveys conducted by nationally recognized consulting and research firms that have focused solely on medium and large companies have found higher rates of benefit offerings compared to those in Table 2.1. For example, other surveys have found that up to 90 percent of medium and large companies offer some form of child care assistance, though the provision of on-site child care remains similarly low in these studies (around 9 percent). Also, 17–31 percent of medium and large companies have reported offering adoption assistance benefits, and 36–38 percent of medium and large companies have been found to offer R&R services.

According to the BLS surveys, employees of larger companies (100+ employees) are much more likely to have access to child care benefits than employees of small and medium-sized companies (25 percent compared with 5 percent in 2007). Nine percent of employees of larger companies have access to on- or off-site child care, compared with 2 percent in smaller firms. Employees of larger firms are also much more likely to have access to R&R services (19 percent versus 3 percent) and an employer subsidy for child care (5 percent versus 1 percent) (BLS, 2006).

Other studies have found much higher rates of employer-provided assistance for child care and other work-life benefits, even as high as 94 percent of employers offering some form of child care assistance and over 50 percent offering alternative work arrangements (Strawn, 2004).

Some employers carefully target child care support, based on productivity data and equity concerns. Bond, Galinsky, and Swanbert (1998) found that employed mothers with children under 13 miss, on average, 6.4 days annually because of family-related issues, including sick children; employed fathers with children under 13 miss, on average, 3.5 days annually for these reasons. Companies with the fewest child care breakdowns employed parents who primarily

**Table 2.1**
**Prevalence of Benefits Supporting Child Care Needs**

| Year | All Workers with Access to Employer Assistance for Child Care (%) | All Workers with Access to Employer-Provided Funds for Child Care (%) | Incidence of On-Site and Off-Site Child Care | All Workers with Access to Adoption Assistance Benefits (%) | R&R |
|---|---|---|---|---|---|
| 1999 | 6 | 4 | 3 | 6 | * |
| 2000 | 4 | 2 | 2 | 5 | * |
| 2001 | * | * | * | * | * |
| 2002 | * | * | * | * | * |
| 2003 | 14[a] | 3 | 5 | 9 | 10 |
| 2004 | 14 | 3 | 5 | 9 | 10 |
| 2005 | 14 | 3 | 5 | 9 | 10 |
| 2006 | 15 | 3 | 5 | 10 | 11 |
| 2007 | 15 | 3 | 5 | 11 | 11 |

SOURCES: BLS, 2004, 2005a, 2005b, 2006, and 2007.

NOTE: "*" indicates that no data were available.

[a] Corrected data.

used center-based or parental child care (Bond, Galinsky, and Swanbert, 1998). Also, researchers have estimated that child care conflicts are the reason for eight to nine absences each year for working parents (Shellenback, 2004).

Union Pacific Railroad uses a program called Rest Easy, which enables the sick children of employees to be cared for by certified nurses at home. While it is not cheap or heavily utilized by employees, anecdotal evidence suggests that the program is greatly valued by both the employees and company management. In addition to local providers such as Rest Easy, a cottage industry of backup care brokers are springing up in metropolitan areas. Lipton Corporate Child Care Centers, for example, offers referrals and priority access to slots in 2,000 Knowledge Learning Corporation child care centers across the country to individual families in need of backup care (Shellenbarger, 2007).

Our interviews revealed that equity, the idea that benefits are distributed in a fair and relatively equal way, is sometimes a concern for employers in providing benefits. The issue of equity, for example, prevented Union Pacific Railroad from actively assisting employees with child care needs for some time, given that most of the employees did not work in a central location. In the end, an on-site center was built at its Omaha headquarters, which employs about 4,000 of their 50,000 workforce. Many employers deal with equity issues by choosing benefits such as R&R services or flexible work schedules that can be used by many people, not just the parents of young children, or by providing targeted benefits that are less costly than providing subsidized on-site child care. As noted above, a popular child care benefit involves access to R&R services to help employees with child care needs. In addition, R&R can be provided to employees regardless of their location. In contrast, companies can provide on-site child care at only a limited number of sites.

In recent years, many companies have moved away from provision of direct child care benefits to provision of benefits that provide employees with more flexibility and choice. This change reflects an increasing recognition by many private-sector companies that a one-size-fits-all approach to benefits generally does not work as well as a more flexible approach.

Also recently, employers have begun to offer benefits of a different kind: changes in the work environment that employees value highly. These types of benefits are highly valued because they enable employees to better meet the varied demands of their work and personal lives. These benefits come in the form of different kinds of financial assistance and policies to allow more flexible work schedules. Bank of America, for example, subsidizes 65 percent of child care costs at its three on-site child care centers. But it also offers employees alternative work arrangements, which include sharing positions, compressed hours, or telecommuting. Flexibility and generous leave policies, combined with financial assistance for child care, are viewed as an appealing package to working mothers ("2006 100 Best Companies," n.d.).

According to Bond et al., 2006, a substantial number of companies reported offering workplace flexibility benefits. The most common flexibility benefit was the freedom to return to work gradually after giving birth or adopting a child (67 percent of companies), followed by being able to take time off work to meet family needs without losing pay (60 percent of companies). Allowing at least some employees to gradually return to work after childbirth or adopting a child increased from 81 percent in 1998 to 85 percent in 2005. One-third of companies allowed employees to periodically alter the start and end times of their daily work schedules, and 13 percent allowed them to make these changes on a daily basis. This benefit was offered by 24 percent of responding employers in 1998 and 31 percent in 2005 (Bond et al., 2006). Table 2.2 describes the availability of flexible benefit plans and other work-life benefits. As shown in the table, employee assistance programs (EAPs), wellness programs, and dependent care reimbursement accounts (DCRAs) are provided by many private employers. Other surveys conducted by nationally recognized consulting and research firms that have focused solely on medium and large companies have found higher rates of benefit offerings, compared with those in shown in Table 2.2. For example, these surveys have found that between 60–86 percent of companies offer flexible spending accounts and 80 percent offer EAPs. These larger employers are also more likely to offer flexible work arrangements (Hattiangadi, 2001).

Table 2.2
**Percentage of All Workers with Access to Other Work-Life Benefits**

| Year | Flexible Benefit Plans[a] | DCRAs | Flexible Workplace Benefits[b] | EAPs | Wellness Programs |
|------|---------------------------|-------|--------------------------------|------|-------------------|
| 1999 | 7 | * | 3 | 33 | 17 |
| 2000 | * | * | 5 | * | 18 |
| 2001 | * | * | * | * | * |
| 2002 | * | * | * | * | * |
| 2003 | * | * | 4 | * | * |
| 2004 | * | * | 4 | * | * |
| 2005 | 17 | 29 | 4 | 40 | 23 |
| 2006 | 17 | 30 | 4 | 40 | 23 |

SOURCES: BLS, 2004, 2005a, 2005b, 2006, and 2007.
NOTE: "*" indicates that no data were available.

[a] For example, Section 125 cafeteria plans.
[b] For example, flextime.

The federal government supports the child care needs of its employees in many of the same ways that private-sector employers do, including access to a dependent care flexible spending account (equivalent to a DCRA), R&R services to local child care resources, on-site child care, child care subsidy programs, and flexible scheduling benefits (GAO, 2007).

## Most Common Private- and Public-Sector Work-Life Benefits

The most common work-life benefits are described below, along with their limitations. Additional benefits not described here include parental education and parental leave policies (e.g., Family and Medical Leave Act policies). It is important to note that many of these benefits have traditionally been related to child care but are being expanded or adapted by private-sector employers to cover all dependents, including elderly and disabled dependents. It is also important to acknowledge the limited applicability to the military compensation system of some of the most popular private-sector benefits, such as flexible work schedules, since they cannot be implemented in the military given the nature of military work.

### Resource and Referral Services

Private-sector employers commonly offer R&R services for child care and other dependent care. This benefit is often provided even when an employer directly offers its own child care services, either on-site or through community-based family child care homes and centers, or offers financial assistance for child care through subsidies or vouchers.

R&R services typically link to a network of child care providers that meet a minimum level of quality, such as being licensed by the state. R&R services are provided to federal employees through a variety of means, including via the U.S. government child care information Web site (http://www.ChildCare.gov). For some employers, the R&R service is a component of their employee assistance program (EAP). Based on BLS surveys, the incidence of employers with EAPs has grown from 33 percent in 1999 to 40 percent in 2006 (BLS, 2006). In the Families and Work Institute's *2005 National Study of Employers,* 66 percent of employers reported providing an EAP (Bond et al., 2006). R&R services assist families in identifying and locating available child care providers. Some services, including the Web site mentioned above, provide information that can help parents select a child care provider that meets their needs and expectations, including information on how to assess quality. However, the R&R services do not address two common child care problems: lack of availability and high costs.

### Flexible Work Schedules

Flexible work schedules have become increasingly popular because they meet a wide range of needs at a minimal cost. According to *Working Mother* magazine's "What Moms Want" survey in 2005 (the number of participants was approximately 600), flextime ranked as the most valuable benefit ("2006 100 Best Companies," 2006). Flexible work schedules may include different start and end times for a work day, compressed schedules, job sharing, telecommuting, flexible leave policies, parental leave (paid leave following birth, adoption, or foster care placement of a child), and part-time work. Flexible work schedules are also available in the federal government; many federal agencies believe that workplace flexibility is an important factor in recruiting and retaining employees with child and adult dependent care needs (GAO, 2007).

## Dependent Care Assistance Plans/Reimbursement Accounts

DCRA is a special type of flexible spending account that employers offer to employees. The accounts allow employees to set aside up to $5,000 in pretax income in a special account. When the employee incurs a valid dependent care expense, he or she can be reimbursed for that expense from the account. The value of the DCRA stems from the tax savings that an employee enjoys from paying for child care with pretax dollars; this benefit is greater for individuals in higher tax brackets. Given its low cost and the relative ease of administering these plans, employers find it appealing. In 2006, BLS data showed that 30 percent of private-industry companies offered employees access to a DCRA (BLS, 2006). These plans are provided in combination with a health care reimbursement account that works similarly to the dependent care account but is used for eligible medical expenses. These plans are offered to all federal employees, but only 7 percent of respondents to a recent survey by the U.S. Government Accountability Office (GAO) said they participate in the plan. The main reasons mentioned for not using the accounts are that they do not pay for care (40 percent), followed by not knowing about the program (26 percent) and not knowing how to use the program (8 percent) (GAO, 2007).

From the employee perspective, the DCRA involves some risks and restrictions. First, an employee must decide how much to set aside at the beginning of the year; this amount can change only if there is a "qualifying event" (such as a marriage, divorce, or birth). Most important, money that is deposited into the account and not claimed for reimbursement is forfeited at the end of each calendar year.

## Voucher or Reimbursement Systems

Under these systems, employees are either given vouchers prior to using child care services, which are then given to child care providers for payment of services, or employees request reimbursement after paying for the care out of pocket. This option gives employees a great amount of flexibility and choice in finding a child care provider, but it does not address availability, quality, or cost issues.

The federal government offers a child care subsidy program to support the needs of employees who work in agencies within the executive branch of government, particularly those who are lower income. The Office of Personnel Management (OPM) instituted final regulations in March 2003 for the federal child care subsidy legislation, "Agency Use of Appropriated Funds for Child Care Costs for Lower Income Employees." The program allows federal agencies to use appropriated funds to assist lower-income employees with the cost of child care for children under the age of 13 or children who are under the age of 18 but are disabled. The children must be cared for in family child care homes or centers that are licensed and/or regulated by state and/or local authorities; there is no requirement for the child care provider to be accredited (OPM, n.d. and 2004).

Federal agencies that participate in the child care subsidy program for their employees have considerable discretion in how they structure their programs. Agencies must annually report to the OPM about their program model, including (1) eligibility and subsidy amount determination; (2) data on the total amount of funds that were disbursed; (3) the number of employees who received the subsidy; (4) the number of children affected; (5) the types and number of child care centers that received the subsidy; (6) the highest, lowest, and average weekly amount of the subsidy; and (7) information about how the agency administered the program and the program's administration costs (OPM, 2004).

Participating agencies have the option of administering the program on their own or paying a subcontractor. The majority of agencies use one of two subcontractors for administering the program, First Financial Associates or the Federal Employee Education & Assistance Fund (FEEA). The roles and responsibilities of these administrative agencies include assisting federal agencies in developing their subsidy program to comply with OPM requirements, collecting and processing applications from employees, verifying license status of providers, invoicing and paying providers on a monthly basis, and performing all regular reporting activities. Contractors usually charge agencies an annual fee per agency ($1,000 for FEEA) and a percentage of the subsidies paid (8 percent for FEEA) (FEEA, n.d.). Each agency has discretion in the way that it structures its subsidy program, including determination of income limits and the amount of subsidy provided. Eligibility determinations are based on an employee's total family income; in most program models, a threshold is set for total family income, and any employee with a total family income below the threshold is eligible for the subsidy. For fiscal year 2006, the income ceilings for participating agencies ranged from $30,000 (Department of Treasury/Bureau of Public Budget) to $70,000 (Central Intelligence Agency).

Program models also vary in the way that the subsidy amount is determined. Some models provide a flat rate for the subsidy, while others use a sliding scale. Some models base the subsidy on the family's total child care costs (a model that is easier to administer), while other models look at the individual cost for each child in care (OPM, 2004).

To participate, employees must apply annually. They must provide tax documents, wage detail, and information about any other sources of child care subsidies they may receive, which will usually reduce their subsidy amount. Parents must also identify the specific child care service they intend to use; they may use the government's R&R services to identify child care services if they have not already done so (OPM, 2004).

Only 2 percent of executive branch employees surveyed participate in the child care subsidy program. When the nonparticipating employees were asked the reason for not using the child care subsidy program, 48 percent said that they do not pay for child care, 33 percent said that they did not know about the programs, 18 percent said that they did not qualify because their total family income was too high, and 11 percent worked in an agency that did not provide the program. Of the employees with total family incomes below $69,000, 37 percent of respondents who did not use the child care subsidy program said that they did not use it because they did not know about it. While it is possible that some of these employees worked in agencies that did not provide the program, these findings demonstrate a need to increase employees' awareness of it (GAO, 2007).

**Off-Site Child Care**

Employers can provide off-site child care in a center that is owned and operated by the employer, or in a center or family child care home in which the employer purchases slots for its employees. If the center is located in a convenient place and open during convenient hours, this can be a great benefit for working parents. The care is usually of relatively high quality because employers seek to avoid liability problems by working with high-quality providers. Employees usually pay a reduced fee for the care.

Some employers, e.g., Texas Instruments, not only offer discounted access to off-site child care, but also contribute to quality enhancements for these subsidized providers, such as for professional development opportunities for the care providers. Texas Instruments contributes to such opportunities through participation in the American Business Collaboration, a group of firms committed to providing access to quality dependent care for their employees.

### On-Site Child Care

As indicated above, on-site child care centers are a rare benefit, even among larger employers, and are almost nonexistent among small employers. Although the idea of on-site child care received substantial attention in the 1990s, and many large employers opened or at least considered opening such centers, interest appears to be waning as telecommuting and work flexibility gain acceptance in the workplace and reduce parental demand for on-site care. These on-site child care centers tend to be of extremely high quality and therefore are extremely expensive to run, particularly if they are open for long hours. Our interview with staff at the magazine *Working Mother* suggested that many of the employers that currently provide on-site child care are reconsidering this option, given the high costs, low usage, and equity issues it raises. Moreover, a 2005 *Working Mother* readership survey found that mothers prefer to utilize child care options near home, rather than near their workplace, if the distance between the two is nontrivial.[2]

In the federal government, there are a number of on-site centers that serve the children of federal employees. As of spring 2007, there were 225 federal (non-DoD) child care centers in the United States. Half of these are operated by the U.S. General Services Administration (GSA) and the other half are operated by other federal agencies. GSA operates child care centers in 31 states, the District of Columbia, and Puerto Rico. While the GSA-operated child care centers are open to the public, half of the slots at each center are reserved for children of federal employees. As of 2005, 58 percent of children cared for in federal child care centers were dependents of federal employees; the federal centers were running at 87 percent of their capacity (the total capacity equals 17,874) (GAO, 2007).

The GSA-affiliated centers are independently operated. Federal agencies pay rent to GSA for use of the child care center space and are issued a revocable license for use of the space. Parent fees and donations typically cover the cost of operating the centers; the subsidy comes in the form of provision of the facility. Parents also form a board of directors to operate the child care centers. All GSA-affiliated centers are accredited by the National Association for the Education of Young Children (NAEYC). These centers are required to have a tuition assistance program, and the amount of assistance available varies by center (GSA, n.d.).

On-site centers raise equity concerns in three ways: first, they are accessible only to employees who work at or near the site where the center is located (usually, corporate headquarters); second, only those who live near their workplace tend to use the centers because many parents are reluctant to transport small children long distances to child care; third, given limits on center size, they are generally used by a very small proportion of employees. Fewer than 10 percent of federal employees who had children in care prior to first grade used a federal child care center (GAO, 2007). The geographic distribution of a federal agency's employees can limit access to federal child care centers. New Mexico provides a good example. The state has 25,000 executive branch employees, but only one-third of those employees work in or near Albuquerque, where the state's single federal child care center is located (GAO, 2007). When most employees of a company work in a single location, such as at universities or in hospital/

---

[2]  Author telephone communication with Tammy Palazzo, Vice President of Research and Women's Initiatives at *Working Mother*, April 19, 2007.

health care systems, the first set of equity issues noted above is not an issue because nearly all employees work at the site where the child care is offered and thus plausibly have access to it. However, the other two equity issues may still apply.

### Other Types of Child Care Services

Included in this type of benefit are backup care, sick child care, summer care, and gap care for periods between school and summer camp, to name a few. These types of care are often needed for unexpected situations; without these benefits, unanticipated child care needs can contribute to absenteeism and reduced productivity.

## Summary

Employers in the private sector and in federal agencies are increasingly shifting to benefits that provide employees with more flexibility and choice. This shift is consistent with what workers report they most value. Employee surveys find that choice and flexibility are highly valued by employees; in many cases, benefits that allow for flexibility are more highly valued than a more costly benefit that limits choice. For child care, this finding suggests that employees generally would prefer a voucher of a lower value that can be used anywhere to a subsidized slot in a child care center at company headquarters.

Providing benefits that offer employees more flexibility and choice requires employers to do a good job of educating employees about their options and places a greater burden on employees to be aware of the benefits and how they work so that they can maximize their value. A Watson Wyatt Worldwide survey (n.d.) described in a 2005 press release found that effective communication about the value of benefits is more important than the value of the benefits themselves. Employees of companies that provided rich benefit package but who did not receive effective communications about the value of their benefits were much less satisfied with them than employees whose companies provided fewer benefits but more effectively communicated with employees about them. Certainly, knowledge about the value of benefits cannot be assumed. In a more recent employee survey, for example, 80 percent of respondents reported that they did not know the value of their benefits (Unum Provident, 2007a and b). When employees do not know the value of their benefits, the utility of those benefits in terms of promoting employer goals of improved recruitment, productivity, and retention is limited.

Regarding child care, it is important to note that while the move to more-flexible benefits appeals to both employers and employees, these flexible benefits do not address two chronic problems in locating and using care: lack of availability and mediocre quality. While vouchers and subsidies may create more demand, these benefits do not address capacity directly. Nor do they help parents recognize quality or necessarily increase the supply of high-quality care. To the extent that reliable child care and high-quality child care are the factors that contribute to improved employee productivity, retention, and recruitment, it is critical that child care benefits support the availability and use of reliable and high-quality care.

But finding and using high-quality care may be difficult. Research consistently shows that the quality of care in this country is mediocre at best (e.g., Peisner-Feinberg et al., 1999). Accreditation by NAEYC is generally understood to be one of the few reliable measures of quality, although states are increasingly developing their own quality rating systems (QRSs) rather than relying on NAEYC accreditation. Zellman et al., n.d., includes a discussion of QRSs in five pioneer states and the role of accreditation in these systems. In the context of our paper, it is crucial to note that these quality measures reflect a specific definition of quality that ignores some issues, such as flexibility, that may be highly relevant to employers and employees.

R&R agencies can address issues of quality, particularly if the state has implemented a QRS that conveys providers' level of quality in an easily communicated way. But R&R agencies often report a lack of high-quality care; this problem is exacerbated when parents have access to quality ratings. Employer vouchers that enable parents to purchase higher-quality (hence, more costly) care may increase demand and supply for such care, but these employer benefits do not address the problems of lack of supply and particularly the short-term lack of supply of high-quality care.

The evidence suggests that child care issues affect productivity and worker morale. Child care problems contribute to absenteeism and reduced productivity; employer efforts to address these problems are associated with more loyal and productive employees. For these reasons, employers have a clear interest in providing some work-life benefits. At the same time, these benefits are costly, and employees have clear preferences, often valuing flexibility over costly but rigid benefits. For these reasons, it is important to provide the right mix of benefits: those that meet employee and employer needs in a context of flexibility and choice.

# CHAPTER THREE

# Description of the Military Child Care System

DoD provides child care for a large number of military families. Two care settings predominate. The first is the CDC, arguably the centerpiece of the military child care system for its large capacity and the fact that the bulk of system subsidies flow to these centers. CDCs provide care for children on a fee-for-service basis during normal working hours, usually 6 a.m.–6 p.m.

The second type is FCC. In FCC homes, individuals (usually, military spouses) who are trained as FCC providers care for up to six children in their own homes.

As of 2001, DoD oversaw 800 CDCs in 300 locations and had relationships with over 9,000 FCC homes that could serve children as young as six weeks and as old as 12 years of age. The total capacity of the CDCs and FCC homes combined was estimated to meet 58 percent of the current need (DoD, 2001).

Eligible families include those headed by a single parent and families with a spouse employed outside the home, or one who is in school or looking for employment (for a limited period of time, such as 60 days). CDC parent fees are based on total family income. FCC providers who accept a subsidy must conform to CDC fee schedules. FCC eligibility and program rules may vary slightly by branch of service.

According to the *Report of the 1st Quadrennial Quality of Life Review,* DoD is currently delivering child care to about 175,000 military children, and about 68 percent of military service members reported satisfaction with their child care services (DoD, 2004a). That report also estimates that there are approximately 900,000 minor dependents of active duty service personnel under age 12 (nearly 500,000 under age 5). In addition, reservists have roughly 400,000 minor dependents under the age of 12.

## History and Background

The DoD child care system is relatively new. It grew out of informal local efforts in the 1970s and 1980s to provide military wives with occasional child care to attend a doctor's appointment or have a quiet lunch or card game with friends. Some installation commanders supported these efforts because they made it easier for spouses to volunteer in a variety of programs on the installation; in many of these instances, the program paid the child care cost for the volunteer.

But the military, like the civilian community, was changing, and more and more spouses were going to work on a regular basis. In addition, the female proportion of the active duty force has been growing since the mid-1970s, as have the number of dual-military families. In

1970, there were 41,479 female active duty personnel (approximately 1 percent of the active duty population). In 2005, there were 202,949 women in the military (15 percent of the active duty population) (DoD, 2005). In terms of family demographics, the 1992 DoD Surveys of Officers and Enlisted Personnel and Military Spouses (DoD, 1997) found that 61 percent of enlisted personnel and 78 percent of officers were married. The most common situation for both enlisted personnel and officers was to be married with a civilian spouse and dependents. Eight percent of enlisted personnel and 8 percent of officers were part of a dual-military couple, and of these couples who had dependents, the majority had children who were younger than six years old (DoD, 1997).

The informal child care programs that were then operating were increasingly unable to meet this growing and changing demand. With varying levels of enthusiasm, the services began to support the development of larger centers that served children for many hours per week.

At the same time, the use of and number of informal in-home child care providers were also expanding. The family-based care that they provided also grew out of informal efforts—in this case, neighbors caring for each other's children so that mothers might have some free time.

But as the number of children in care grew, the informal, largely unregulated network of care began to show signs of stress. Waiting lists increased, leading to concerns about the availability of care. Incidents of child abuse in several CDCs in the late 1980s raised questions about the quality of care. Abuse incidents at the Presidio Child Care Center prompted the congressional member for that district, Barbara Boxer, to ask Beverly Byron, the chairperson of the House Subcommittee on Military Personnel and Compensation, to hold hearings regarding the circumstances that may have allowed these child abuse incidents to occur.

One of the first issues that emerged in the hearings was high staff turnover because of very low caregiver wages. Another issue that emerged from the hearings was substantial variability across services in the way that child care was operated and managed, including variation in the level of appropriated funds support and the rigor of inspection programs. The hearings also underlined the inadequacy of financial support for military child care. Congress determined that a subsidy was needed to ensure that adequate funds were available to deliver high-quality care.

MCCA was Congress's response to these concerns. MCCA sought to improve the quantity and quality of child care provided on military installations. An additional aim of the act was to standardize the delivery, quality, and cost of care across installations and military services, which in 1989 varied considerably.

MCCA relied heavily on four policies to realize the key goals of the legislation: substantial pay increases for those who worked directly with children, with pay raises tied to the completion of training milestones; the hiring of a training and curriculum specialist in each CDC to direct and oversee staff training and curriculum development; the requirement that parent fees (which would henceforth be based on family income) be matched, dollar for dollar, with appropriated funds; and the institution of unannounced inspections of child development centers to be conducted four times yearly. The legislation specified a series of remedies for violations discovered during inspections. It also provided for the establishment of a child abuse reporting hotline.

The framers of MCCA were primarily concerned about protecting children in DoD's care, an understandable focus given that a child abuse allegation had precipitated the legislation. They did not consider at that time whether the system that had grown up informally and organically to meet local needs was the best way to provide what would, with the impetus from MCCA, become a significant employee benefit. Indeed, no consideration was given to whether supplying child care through CDCs and FCC was the best way—for DoD, military parents, or their children—to supply military child care.

This short history illustrates an important contrast between employer-sponsored child care in the private sector and in DoD. While private-sector firms have developed family support benefits to assist employees in meeting the demands of their work and personal lives, they typically choose benefits based on documented evidence that there is a "business case" for these work-life benefits. They monitor these benefits and modify them over time to ensure that the benefits are contributing to their own goals, typically improved recruitment, productivity, and retention. In contrast, the military child care system began informally without support or input from the employer (DoD). This organic system was then formalized and expanded, against DoD opposition,[1] by MCCA. Only in recent years have new options emerged, as discussed below. But even these new options have not been subjected to a rigorous assessment in terms of what they accomplish for the military and the families using them; nor have they become part of a more general discussion of the assumptions underlying them and the child care system as a whole.

## Overview of the Current Military Child Care System

DoD currently offers a range of child care programs for military families, either through direct provision or contracts with third parties. DoD oversees CDCs, FCC, and school-age care programs (SACs), as well as the R&R system, Military OneSource, whose Web site provides referrals to the child care system, while a civilian supplier, NACCRRA operates military child care subsidy programs (see **Military Child Care Subsidy Programs,** below). In its entirety, DoD's child care program is the largest employer-sponsored child care program in the country, which is well-known for consistency and high quality (93 percent of DoD CDCs are accredited by NAEYC). DoD's child care assistance is viewed and operated as a family support service and is made available on a first-come, first-served basis to families that apply for services, with priority given for highly coveted CDC spaces to family types (single parent, dual military) perceived to be most in need of care in order to report for duty.

Although DoD-wide policy guidance and MCCA establish broad parameters within which the military child care system must operate, the system reserves a fair amount of discretion for installation commanders and the services. In particular, the decision to request resources to build a CDC is made by the installation commander, who must weigh the relative benefits of a new CDC against, for example, a new airfield, a new fitness facility, or a new dining hall. The services also have funds that can be allocated for new construction, but there is competition for these resources. Although MCCA requires that DoD contribute appropriated funds to CDC operation in an amount at least equal to the amount received in parent fees, installation commanders are responsible for determining the extent to which base

---

[1] MCCA was initially opposed by DoD and by all the services, generally on the grounds that the problems in the system had been identified and would be addressed through a new DoD Instruction that was then being written and which was published in March 1989.

resources will be used to support the operation of child care activities, such as subsidies for FCC, support for extended hours of child care, or support for child care for families with special needs. Specific policies and procedures vary by installation as well, such as hours of operation and the number of hours per day that a child can stay in a CDC. As a result, there is quite a bit of variation in child care at the local level within DoD.

**Child Development Centers**

DoD's CDCs provide child care services to military service members' children from the age of six weeks to six years. There are about 800 CDCs at over 300 locations across installations throughout the world. Operating hours are from approximately 6 a.m. to 6 p.m., Monday through Friday. CDCs are managed through each service—the U.S. Army, the U.S. Navy, the U.S. Marine Corps, and the U.S. Air Force. CDC fees are assessed on a sliding scale, based on total family income, in an effort to promote affordability. Low-income families can pay as little as $43 a week for 50 hours of child care, while those with high incomes may pay as much as $126 a week. These fees are low compared to the amount that most American families pay for child care. According to 2002 Census data, U.S. families with working mothers (non-self-employed) paid an average of $96 a week for child care. Families with younger children, under age 5, paid twice as much on average for weekly child care ($122), compared with families with children ages 5–14 ($60 a week), and the payment represented a larger proportion of their income (10 percent compared to 4 percent). Low-income families with working mothers paid an average of $67 per week, compared with $98 per week for families not in poverty (Overturf Johnson, 2005).[2]

CDCs are open to children of active duty military members, civilians employed by DoD, and military retirees. In order to use CDC care, a military member's spouse must be employed or attending school (as described above).

Nearly all DoD CDCs are enrolled to capacity and have waiting lists. These waiting lists may be quite long, particularly for infant, pretoddler, and toddler care. Most CDCs allocate available slots first to the children of single parents and dual-military parents, then to children of other active duty families. DoD civilians and military retirees, while also eligible to use CDC care, receive the lowest priority on the waiting lists. This priority reflects an effort to address readiness issues through the allocation of CDC care, but in reality, center-based care, with its limited hours of operation, cannot meet the needs of all single-parent and dual-military families. Many believe that FCC, which is inherently more flexible, is better suited to families that have limited control over their working hours.

**Family Child Care**

For those who prefer a more intimate child care setting or are on waiting lists for CDCs, DoD certifies individuals to offer child care in their own homes and provides support for such FCC. Some of these providers are available for extended hours and weekends, when CDCs are

---

2  In considering these cost differences, it is important to keep in mind that nearly all military child care centers are accredited (a measure of high quality), whereas fewer than 10 percent of centers nationwide meet accreditation standards. One of the largest expense categories for child care centers is personnel costs, and this is especially true for high-quality child care. In 2000, the national average salary for family child care providers was $4.82 an hour and $9.43 an hour for preschool teachers, while the average starting salary offered in ten exemplary child care programs was $11.82 in 2002–2003. Parent fees typically cover the largest proportion of child care costs. Therefore, the cost of improving child care quality is often shouldered by parents (Greenberg, 2007).

closed. There are about 9,000 FCC providers located on or near bases. FCC providers must go through DoD background checks and be licensed by the state in which they operate. They are also encouraged to achieve accreditation by the National Association of Family Child Care, in addition to DoD and service requirements.

That FCC was originally conceived as a spouse employment program explains some of its policies, notably the right of providers to set their own fees. Over the years, some installations have begun to subsidize FCC programs, although the subsidy levels (usually for insurance and equipment) are very modest compared with those offered in CDCs. On some installations, providers are offered targeted subsidies, e.g., to provide care for infants and special needs children. When a CDC at a particular installation decides to stop providing care for infants because the costs are so high relative to care for older children, subsidies for infant care in FCC are sometimes begun.[3] If an FCC provider receives any subsidy, she must then use the CDC fee policy in setting her fees for care. Otherwise, the fees charged by FCC providers are negotiated directly between the provider and the parent.

### Care for School-Aged Children

DoD also offers care for children ages 6 to 12 before and after school and during holidays and summer vacations. These programs are designed to complement education that children receive in schools by emphasizing community needs, family values, and overall development. The programs are often located in youth centers, FCC homes, and other appropriate facilities. In most cases, the programs provide formal care to the youngest of school-aged children. Older children simply use youth center facilities at a much-reduced cost (with much less supervision).

### Resource and Referral

Military OneSource is a 24-hour R&R service provided by DoD for all active duty, Guard, and Reserve members and families. Trained consultants are available to provide information and referrals on a wide range of work-life topics, from child care to finances. For child care, the Military OneSource Web site features information on NACCRRA military child care programs and direct links to application forms.

### Military Child Care Subsidy Programs

DoD partners with NACCRRA to administer a number of subsidy programs designed to help military families pay for child care:

- Operation: Military Child Care
- Military Child Care in Your Neighborhood
- Child Care Support for Severely Injured Military Members.[4]

NACCRRA takes the lead on processing applications and allocating subsidies on DoD's behalf. The programs provide subsidies to activated or deployed Guard and Reserve members, deployed active-duty personnel, and service members who live in areas where on-base care is not available. The programs also provide special assistance to military members who are

---

[3]  As noted above, the DoD fee policy is tied to total family income, not to the cost of care. Since the cost of delivering care is much higher for infants, a decision not to serve infants in a CDC is an efficient way for the CDC to save money.

[4]  For more information on these programs, see NACCRRA, n.d.

injured in the line of duty. Eligible members receive help in identifying providers that meet DoD quality standards and in paying for child care through fee assistance. Eligibility and the level of assistance are determined by family income, circumstances, geographic location, military child care fee policies, and available funding.

**Child Care Use by Military Families**

A 2004 survey of military families with children under age 12 provides a clear description of how military families are caring for their children (Gates, Zellman, and Moini, 2006). Because the child care subsidy programs described above are relatively new, the survey does not address the use of these programs. Several points are worth emphasizing. First, a small fraction of all military families use child care provided through the military child care system. Second, different types of military families—single parent, dual-military parents, and families with a civilian spouse (either employed or not)—use DoD-sponsored child care at different rates (see Figures 3.1–3.3). These rates are explained by differences in the use of parental care by families and by priorities for access set by the services. Third, other factors, such as family income and where a family lives, are related to child care use.

**Figure 3.1**
**Child Care Use for Preschool-Aged Children—Single-Parent Military Families**

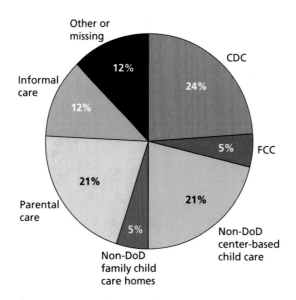

SOURCE: Gates, Zellman, and Moini, 2006.
NOTES: The number of single military parents in the study sample was 58. Although it might seem surprising that so many single military parents use parental care, our analysis of survey responses revealed that most of these are families headed by an unmarried male military member who reported that care is provided by the child's mother. We surmise that these are families in which the mother, who is not married to the father, lives near or with the military member so that she is able to provide care.

RAND *OP217-3.1*

**Figure 3.2**
**Child Care Use for Preschool-Aged Children—Dual-Military Families**

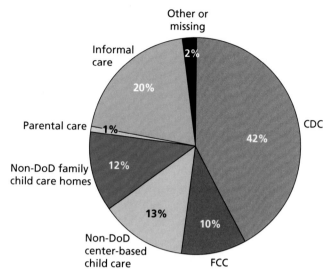

SOURCE: Gates, Zellman, and Moini, 2006.
NOTE: The number of dual military families in the study sample was 241.

**RAND** *OP217-3.2*


**Figure 3.3**
**Child Care Use for Preschool-Aged Children—Families with a Civilian Spouse**

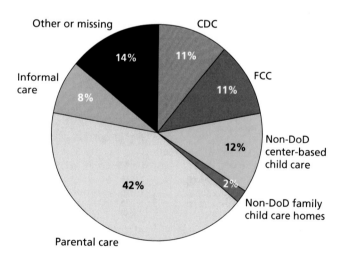

SOURCE: Gates, Zellman, and Moini, 2006.
NOTE: The number of families with a civilian spouse in the study sample was 162.

**RAND** *OP217-3.3*

**What Care is Used by Military Families?** The DoD child care system serves a relatively small fraction of military families with children overall, and use varies dramatically by child age and by family type.

Using only their own survey data, Gates, Zellman, and Moini (2006) found that about 16 percent of survey respondents with preschool-aged children use a CDC as the primary form of care for their child, but the figure is 42 percent for dual-military families and 24 percent for single-parent families.[5] Only 11 percent of families with a civilian spouse use a CDC for their preschool-aged children. Overall, about 10 percent of military families use DoD-sponsored FCC, but only 5 percent of single-parent families do. The fraction of families with preschool-aged children who use a civilian child care center (13 percent) is only slightly lower than the fraction who use a DoD-sponsored CDC.

Among families with school-aged children, just over 10 percent use any DoD-sponsored option (including CDCs, FCC, SACs, or youth centers). Again, dual-military families are the heaviest users of DoD-sponsored programs—about 20 percent use DoD programs for their school-aged children. The figures are 14 percent for single-parent families and 9 percent for families with a civilian spouse. Among dual-military families, 64 percent rely on a formal civilian option (primarily civilian after-school programs). See Figures 3.4, 3.5, and 3.6.

**Figure 3.4**
**Child Care Use for School-Aged Children—Single-Parent Military Families**

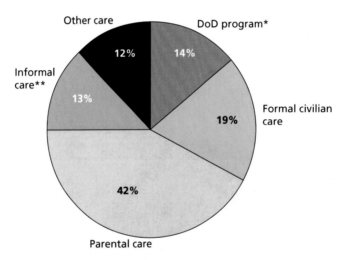

SOURCE: Gates, Zellman, and Moini, 2006.
NOTE: The number of single parents in the survey sample was 53.
* DoD programs for school-aged children include CDC, FCC, SAC, or Youth Center.
** Informal care for school-aged children includes care provided by siblings, relatives, or friends, and by the child him or herself. While DoD regulations prohibit self-care by children 12 and under, we suspected that it occurs, and with assurances of confidentiality in place, we asked about it.

**RAND** *OP217-3.4*

---

[5] The survey asked parents to report on the type of care used for the most hours per week to care for children while the parent or parents were working or going to school. Parental care (care provided by the mother or father) was an option that could be selected. The survey also asked families to report on secondary care arrangements. So a military spouse who cares for her children full time but periodically uses babysitters would report "parental" care as the primary form of care and "informal" care as a secondary form. In this paper, we present information on only the primary form of care used. Information on use of secondary child care arrangements can be found in Gates, Zellman, and Moini, 2006.

**Figure 3.5**
**Child Care Use for School-Aged Children—Dual-Military Families**

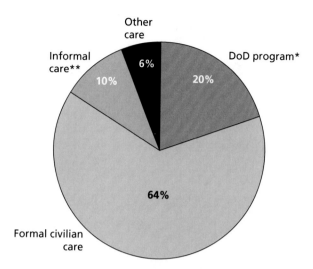

SOURCE: Gates, Zellman, and Moini, 2006.
NOTE: The number of dual-military families in the study sample was 161.
* DoD programs for school-aged children include CDC, FCC, SAC, or
Youth Center.
** Informal care for school-aged children includes care provided by
siblings, relatives, or friends, and by the child him or herself.  While DoD
regulations prohibit self-care by children 12 and under, we suspected
that it occurs, and with assurances of confidentiality in place, we
asked about it.

RAND *OP217-3.5*

Gates, Zellman, and Moini (2006) survey data reveal that roughly half of military families use a formal child care arrangement (i.e., a center, a family day care home, or an after-school program) to care for their children. For families that do not use formal child care, parental care is the most common form of care used. Use of parental care varies by child age and by family type. Overall, for one-third of families with preschool-aged children and 46 percent of families with school-aged children, parents are the only child care providers. Almost no dual-military families use parental care for either preschool- or school-aged children. In contrast, 21 percent of single-parent families use parental care for preschool-aged children, and 42 percent use it for school-aged children. Survey responses suggest that this care is being provided by a custodial parent who is not currently married to the military member. For families with a civilian spouse, the figures are 42 percent and 50 percent, respectively.

Among families that use some form of child care other than parental care, the specific option selected varies in systematic ways across families. Across the board, families living off base are less likely to use DoD-sponsored child care options, and the tendency to use civilian options increases with the distance between home and the installation. This finding suggests that proximity to home is an important characteristic of child care and that the DoD-sponsored child care, which is typically located on base, is a less attractive option for families that live further from the base. Family income also appears to play a major factor in child care choice, especially for families with preschool-aged children. Families that earn more than $75,000 per year are much less likely to use a CDC and are relatively more likely to use FCC.

**Figure 3.6**
**Child Care Use for School-Aged Children—Families with a Civilian Spouse**

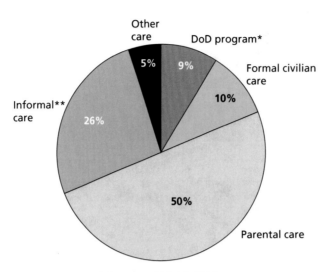

SOURCE: Gates, Zellman, and Moini, 2006.
NOTE: The number of participants in the study was 198.
* DoD programs for school-aged children include CDC, FCC, SAC, or Youth Center.
** Informal care for school-aged children includes care provided by siblings, relatives, or friends, and by the child him or herself.  While DoD regulations prohibit self-care by children 12 and under, we suspected that it occurs, and with assurances of confidentiality in place, we asked about it.

RAND OP217-3.6

For school-aged children, families of military officers are more likely to use parental care as are families that live in areas with relatively high median incomes.

**What Fraction of Military Members Are Served by the Military Child Care System?** The preceding discussion summarizes the type of child care used by military families with children under age 12. Overall, the discussion indicates that a small fraction of military families with children are using a form of DoD-provided child care. Viewed from another perspective, roughly half of families with children under age six who use something other than parental care as their primary child care type are using DoD-sponsored CDCs or FCC and about 20 percent of military families with school-aged children that use something other than parental care use DoD programs.

In considering child care from a compensation perspective, it is also worthwhile to examine the fraction of all military personnel who are currently receiving the benefit. There are no readily available data on the percentage of families served by the military child care system, but the survey data discussed above can be combined with information on the family status of the military population to construct rough estimates. In particular, we focus on estimating the fraction of military members who are served by CDCs and FCC. Using published data from the 2003 Status of Forces Survey, which provides information on the number of families with children under age 12, we assumed conservatively that *all* families with a child under age 12 have a child under age 6. We then calculated the percentages of military members who have children under age 6 and who fall into specific family type categories: married to a civilian spouse, dual military, or single parent, as shown in Table 3.1.[6]

To calculate the fraction of military members who are served by a particular form of child care, we multiply the percentages in Table 3.1 by the percentage of military members of that type who use the particular form of child care. For example, Figure 3.2 suggests that 42 percent of dual-military families with children under six years of age use CDC care. Dual-military members with children under six represent at most 5 percent of all military members. Therefore, we estimate that at most, 2.1 percent of military members are dual-military members with children under six who use CDC care. We perform similar calculations for other family types and find that at most, 3.6 percent of military members are married to a civilian, have children under six, and are CDC users and at most 1.2 percent of military members are in single-parent families with children under six and use a CDC. Adding these percentages across family types, we estimate that the CDCs are serving at most 7 percent of all military members at a given time. Using the same procedure, we estimate that at most 4 percent of military members are using FCC at a given time.

**Unmet Needs for Child Care.** The RAND child care survey asked families that were not using a formal child care arrangement whether they had an unmet need for child care in a formal arrangement.[7] Overall, 9 percent of military families expressed an unmet need for child care. Here again, dual-military families look different than other families. Dual-military families are very unlikely to have an unmet need for care for preschool-aged children, but over 10 percent express an unmet need for care for school-aged children. Among families with a civilian spouse, over 10 percent have an unmet need for care for preschool-aged children. The unmet need for care for school-aged children is very low (3 percent) for families with a non-working civilian spouse, but much higher (15 percent) for families with a civilian working

**Table 3.1**
**Percentage of Military Members with Children Under Age Six, by Family Type**

| Family Type | Percentage |
| --- | --- |
| Married to a civilian spouse plus children under six | 33[a] |
| Dual military plus children under six | 5[b] |
| Single parent plus children under six | 5 |
| No children under six (all family types) | 57 |

[a] This is percentage of all military members who have the listed family type and children under six.

[b] This figure represents about 2.5 percent of military families since each dual-military member counts twice.

---

[6] This conservative estimate allows us to estimate an upper bound. While it is likely that some families with a child under 12 do not have a child under age 6, it is impossible for families that do not have a child under age 12 to have a child under age 6. We were not able to find information specifically on the percentage of military families that have children under age 6.

[7] As defined in Gates, Zellman, and Moini (2006), *unmet need* refers to families that report that they would like to use a formal child care arrangement but are not doing so, regardless of the reason why. That report provides a detailed analysis of unmet need.

spouse. Among single parents, the probability of having an unmet need is 8 percent for families with preschool-aged children and 6 percent for those with school-aged children. Overall, then, it appears that having an unmet need is greatest among families with two working parents and school-aged children, as well as families with an employed civilian spouse and preschool-aged children.

**Unmet Preference for Particular Child Care Options.** Most CDCs have waiting lists; these lists are often pointed to as evidence that there is enormous excess demand for DoD-sponsored child care (i.e., CDC spaces provided at the current price). However, it is not clear whether families are using the child care options that they most prefer or whether limited availability, lopsided subsidies, or other factors are leading families to feel that they cannot access the type of care they really want for their children. Based on the results of Gates, Zellman, and Moini's (2006) military child care survey, which was fielded in 2004, there does appear to be excess demand for subsidized CDC care, but at the same time there are individuals who are using DoD care, even CDC care, who would prefer to use something else. The survey asked families if the child care option they were currently using was their most preferred option. Twenty-nine percent of families with preschool-aged children and 19 percent of families with school-aged children reported that they would prefer a different option from the one they were using. This is consistent with the finding from the first quadrennial quality of life review (DoD, 2004a), which reports that about 68 percent of military service members are satisfied with their child care services. Among families with preschool-aged children that expressed unmet preferences, over half would prefer to be using a CDC. Among families with school-aged children, about 20 percent who expressed unmet preferences would like to be using a DoD after-school program. Dual-military families were far more likely to express unmet preferences for school-aged children. This finding may reflect a strong preference among dual-military families for DoD care and a relative lack of such care for school-aged children.

The families that are not using CDC care but who would like to are balanced by a nontrivial number of families that are using DoD-provided care but would prefer to use a different option. Survey data reveal that between one-quarter and one-half of families using DoD care would prefer some other type of care.

## Summary

The survey results described above reveal that the military child care system is just one of many options that military families consider when determining how to care for their children. The survey also reveals very different outcomes for different types of families: both in terms of the type of care used and in terms of the availability of care they would like to use.

Military families with a civilian spouse are much more likely to use parental care, and are consequently less likely to use any formal child care option, including DoD-sponsored child care. Focus groups conducted as part of the survey project indicate that for many families, parental care is the most preferred type of child care (Moini, Zellman, and Gates, 2006). Most of these families (just under 70 percent) that use parental care have a civilian spouse who does not work outside the home. However, some military families with two working spouses reported that they take advantage of flexible work hours or staggered shifts in order to care for their children themselves. Families that use parental care may not incur an explicit child care expenditure, although the forgone income of a nonworking civilian spouse should be considered as an implicit cost of using this arrangement. It is also important to point out that half the survey respondents expressed the opinion that work opportunities for civilian spouses were limited in the local area and that good opportunities required a long commute. These families may be using parental care because of limited employment opportunities for the spouse. Finally, it is worth emphasizing that families with a civilian spouse are more likely to express an unmet need for child care—particularly for preschool-aged children.

Dual-military families have a much different child care experience than other families. Parental care is not a realistic option for dual-military families, and these families are the most likely to use the DoD child care system—particularly DoD CDCs. This finding is not surprising given the fact that dual-military families typically receive preference in CDCs. The system appears to be reasonably successful in meeting the needs of dual-military families with preschool-aged children. They are the least likely to express unmet child care preferences. However, dual-military families are most likely to express unmet preferences for care for school-aged children. Specifically, dual-military families seem to prefer DoD-sponsored after-school programs, but often lack access to them. One reason may be that at least some of these programs do not offer transportation from school to the program site.

Single-parent families seem to fall between dual-military families and families with civilian spouses in terms of their use and preferences. We had initially expected single-parent families to look more like dual-military families, under the assumption that if a child is living with a single military parent, the other parent is out of the picture and unavailable to provide child care. The preference single military parents receive on waiting lists for DoD-sponsored care at most military installations is also based on this assumption. However, this assumption appears to be invalid; a fair number of single (male) military members reported that care was provided by the child's mother, a finding that suggests that at least some single military parents do have other child care options available to them and are not as reliant on the DoD system as is assumed.

In terms of an employee benefit, military child care is not reaching a large fraction of the total military population. At most, 7 percent of military members are served by CDCs and another 4 percent by FCC homes. Even among families with children under age six, fewer than half use DoD-sponsored child care. In the next chapter, we look at what DoD spends in order to provide those benefits and then consider what value DoD may be getting for these expenditures.

# CHAPTER FOUR

# Military Child Care from the Perspective of DoD as an Employer

In assessing the military child care system from an employer perspective, DoD must consider not only whom the system is serving, but also the costs to DoD of providing the benefits and the value that DoD receives in return.

## Cost of DoD-Sponsored Child Care

Singer and Davis (2007) estimate that DoD spends about $480 million annually on military child care. Given the wide variety of child care options used by military families and the factors that are related to child care use, it is worthwhile to consider the cost drivers of the military child care system. To do this we draw on information from Zellman and Gates (2002). That study relied on the results of a survey of 69 military installations to construct estimates of the cost of different child care options. The survey was conducted in the fall of 1999 and asked installations to provide information from fiscal year 1998. The survey data were analyzed to assess the total cost of different child care options and to assess the pattern of subsidies from DoD to military parents.

In discussing the results of the survey below, we adjust the dollar figures using the Consumer Price Index to reflect 2007 purchasing power. This method is likely to understate the actual increase in the cost of different child care options.[1] However, this detailed cost survey is the most current source of information on the costs of operating the child care system, and the data provide a useful starting point for discussion.

### Cost

In analyzing the cost of military child care, Zellman and Gates (2002) considered the amount of money that DoD and parents together spend on different types of child care. The authors used information from the cost survey to construct estimates by care type (CDC, FCC, and contractor-provided center care) and by child age category.[2] The analysis did not account for facility construction costs. In the case of both CDC care and contractor-provided center care,

---

[1]  In 2003, the *Wall Street Journal* noted that the cost of child care had been increasing at a faster clip than most other consumer costs—about 5 percent per year. This increase may be due, in part, to increases in the average quality of care. See *Wall Street Journal,* n.d.

[2]  MCCA (as well as most state licensing and NAEYC accreditation criteria) stipulates specific staff-child ratios by age group. As a result, child age is a key driver of the cost of center-based child care. The age categories are defined as follows: infant—0–11 months; pretoddler—12–23 months; toddler—24–35 months; preschooler—36 months–five years; school age—five years and up.

those costs are borne by DoD. A 1999 GAO study of U.S. Air Force child care estimated these facility costs at about 10 percent of the total. For FCC, the care is typically provided in military housing (typically DoD or government-leased housing), and facility costs are not accounted for directly.

Figure 4.1 summarizes the cost of DoD child care options as estimated by Zellman and Gates (2002).[3]

In general, the cost of CDC care is substantially higher than the cost of FCC or the cost (i.e., the price charged) to DoD and parents for child care provided to DoD by civilian contractors. The difference is greatest for children under two years of age (infants and pretoddlers), and less so for older children. In fact, the cost of care for preschool-aged and school-aged children provided in DoD centers is sometimes lower than the price charged to DoD and parents for contractor-provided care. FCC is less costly to provide than CDC care; the difference is estimated to be over $9,000 per year for each infant, and about $1,600 per year for each school-aged child. DoD has expanded support for FCC in order to increase both the capacity and the quality of FCC. As such, it is likely that the direct and indirect costs of FCC have increased relative to the costs of CDC care. Nevertheless, FCC likely remains a far less costly option for providing care to the youngest children.

**Figure 4.1**
**Estimated Annual Cost of Child Care, by Child Age Group**

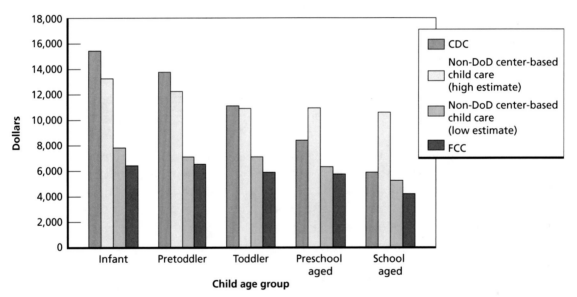

SOURCE: Zellman and Gates, 2002.
NOTE: Dollar amounts have been revised to reflect 2007 dollars.

RAND OP217-4.1

---

[3]  These average costs reflect straight, unadjusted averages for a sample of installations that were included in the study. The sampling strategy and analytical methods are described in Zellman and Gates (2002). Installations were stratified in terms of whether they were located in "rich" (i.e. high-cost-of-living) areas and remote areas, and the distribution of responding installations across the four categories was similar to the distribution of all installations. The authors show that CDC costs are higher at installations that are located in high-cost-of-living areas, and they are higher at installations that have CDCs with a smaller average size. The authors also found that FCC costs were higher in areas with higher costs of living.

**Subsidy**

A vast majority of the child care resources spent by DoD is devoted to care provided in CDCs. Although parents do pay fees to the CDCs, these fees cover less than half of the cost of operating CDCs.[4] Parent fees for CDC care are set at the installation level with guidance from the services and DoD based on MCCA. The fee schedules vary by family income. Families with higher incomes pay higher fees.[5] Parent fees do not vary by child age, even though the cost of providing care varies substantially by child age, as shown in Figure 4.1. Zellman and Gates (2002) used information on average income Category III parent fees along with the cost estimates generated from a 1999 cost survey to estimate the fraction of total CDC costs covered by parent fees. We have updated the estimates contained in that report to reflect the value of the subsidy in 2007 dollars. To do this, we assume that the subsidy rate by child age has remained constant over time and multiply that rate by the estimated cost of providing child care in 2007 dollars. Figure 4.2 displays the estimated value of the subsidy by child age group.

Figure 4.2 illustrates that the DoD subsidy is worth nearly $11,000 per year to parents of infants who are in income Category III. That value declines by child age to about $1,400 per year for income Category III parents of school-aged children. For each child age level, the value of the subsidy would be higher for lower income category parents (Categories I and II) and lower for high-income category parents. Indeed, a high-income parent of an older child in the CDCs may be paying DoD more than the actual cost of caring for his or her child.

In 1999, few installations (14 out of 69) were providing any direct fee subsidies for FCC. Of those few, many were providing subsidies only for infant care in FCC homes. The maximum subsidy provided was $90 per week for each child. In 2007 dollars, that would amount to a subsidy of $114 per week, or nearly $5,000 per year per child. Most installations that did provide subsidies provided far less than this amount.

We did not examine the cost of care provided to school-aged children, although we have reason to believe that the subsidies for these children are not large.

## Contributions of the Military Child Care System to Outcomes of Interest to DoD

A crucial question for DoD to consider is whether the system as it is currently structured is providing DoD with value in terms of improved recruiting, readiness, and retention. As discussed in Chapter Two, private-sector employers typically provide child care benefits to employees for several key reasons: as a recruiting, retention, and productivity promoting tool; to assist employees in meeting the special demands of employment, such as extended hours and

---

[4]   MCCA requires that each dollar in fees spent by parents be matched with one dollar of support from appropriated funds, which sets the subsidy floor at 50 percent overall. Typically, installations pick up costs associated with building maintenance, purchase of high-dollar equipment, and janitorial services, increasing the overall subsidy rate.

[5]   For the 2006–2007 school year, there were six income categories. Category I: $0–$28,000; Category II: $28,001–$34,000; Category III: $34,001–44,000; Category IV: $44,001–55,000; Category V: $55,001–70,000; Category VI: over $70,000. The income category is determined by total family income. In 1998, DoD had only five income categories. Income Category III (used to calculate subsidy rates) included families with total family income between $34,000 and $44,000 (in FY 1998 dollars).

**Figure 4.2**
**Estimated Value of the CDC Subsidy Paid by DoD per Child, by Child Age Group for Income Category III Parents**

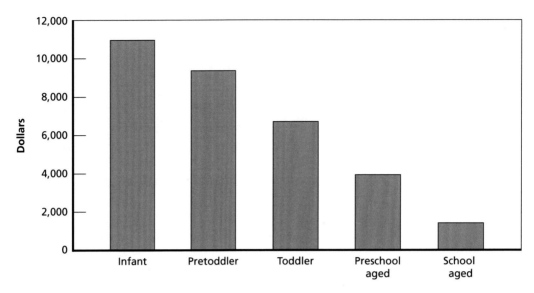

SOURCE: Zellman and Gates, 2002.
NOTE: Dollar amounts have been revised to reflect 2007 dollars.

**RAND** *OP217-4.2*

shift work; and to assist employees in balancing the demands of their jobs with the demands of their personal lives.

As mentioned in the introduction to this paper, evaluations of the military child care system since the implementation of MCCA have tended to focus either on measuring the number of spaces provided by the system or on assessing the quality of care provided. To our knowledge, there has been no systematic evaluation of the relationship between military child care and readiness, retention, or recruiting. In this section, we discuss what is known about the extent to which the military child care system is contributing to these DoD goals.

**Readiness and Productivity**

The 2004 RAND child care survey by Moini, Zellman, and Gates (2006) asked military families whether child care issues had prevented a parent from reporting for military duty either following the birth of a child or after the most recent relocation. Such problems were an issue for over one-third (36 percent) of dual-military families and over 10 percent of single parents, but only for fewer than 1 percent of families with a civilian spouse. The survey also asked whether child care issues forced a parent to miss work or be late to work in the past month. Over half (51 percent) of military mothers and 22 percent of military fathers reported being late to work in the last month because of child care issues. Similarly, over two-thirds (37 percent) of military mothers and 7 percent of military fathers reported having to miss work because of child care issues. Clearly, female military parents are carrying a greater child care burden and covering for child care inadequacies more often than their male counterparts. This situation is mirrored in the experiences of private-sector employees as well. These survey results suggest that child care issues have implications for the productivity of military members and that this is true even for families that are using DoD-sponsored child care options.

Absenteeism and readiness may be influenced by the flexibility of child care arrangements that families use. One measure of how well current child care options are meeting the needs of military families is the degree to which families using a particular type of care need to use more than one arrangement in order to care for their children in the context of work. Moini, Zellman, and Gates (2006) found that about 20 percent of families surveyed reported using more than one arrangement in the past week. For a vast majority of these families, the secondary arrangement was an informal one (such as a friend, nanny, or relative), and it was used for fewer than ten hours per week. The authors examined the fraction of families that used a secondary arrangement by family type and by the type of primary care used; the data reveal that the patterns differ dramatically by family type. Dual-military families are the most likely to use a secondary arrangement (nearly half reported doing so). However, whereas nearly all dual-military families that use FCC or informal care options for preschool-aged children reported using a second option, only 15 percent of CDC users did. Overall, primary reliance on informal child care options (such as relatives, nannies, or friends) is more likely to require parents to find supplementary child care. In contrast, few two-parent families that use a CDC have to supplement that care with another form of child care. In that sense, the CDCs appear to be meeting the needs of the families that are using that type of care, although we do not know the extent to which some families choose not to use a CDC because it would not meet their needs as a result of their work schedules.

**Retention**

Given the challenges of juggling child care and military work, DoD may be concerned about the effect of child care on retention. Although there is no direct evidence on this subject because DoD does not systematically track who uses military child care, the 2004 RAND child care survey did ask families to report on the likelihood that a military member would leave the military because of child care issues. The survey found that families with preschool-aged children were much more likely to report such a propensity. Among families with preschool-aged children, dual-military families were 30 percentage points more likely to report a propensity to leave the military than single military parents because of child care issues. Surprisingly, the analysis of survey results also revealed that, controlling for other characteristics such as family type, CDC users were also more likely to report a propensity to leave the military because of child care issues. This latter finding is troublesome from an employer perspective and may reflect the inflexibility of this type of care. DoD is devoting a substantial portion of child care funds to the CDCs, yet appears to be reaping few benefits.

**Recruitment**

We are aware of no studies that have examined the implications of military child care for recruiting. Furthermore, we were not able to identify any military recruiting efforts that specifically highlight the military child care system.

**Who Receives the DoD Child Care Subsidy (and Are They Aware of It)?**

Parents of young children who are in low-income categories and use the CDC receive child care subsidies worth over $10,000 per year. Other CDC families also receive significant subsidies, although the value of the DoD subsidy is lower for parents in higher income categories and for parents of older children. Families using other DoD care, such as FCC, receive little to no subsidy.

Applying the survey results discussed above concerning the number of military parents to these subsidy data, we can see how limited these child care subsidies actually are. About 33 percent of military members have children under 12 years of age. Slightly more than half of the children under 12 are between the ages of 0 and 5. However, only about 16 percent of these parents use CDC care. This finding suggests that most military families with children do not receive any child care benefit from DoD, while those who use CDCs, particularly those who have infants being cared for in CDCs, receive a benefit equivalent to as much as one-third of their basic pay.

Among those who do receive a benefit from DoD, many are unaware of the subsidy since they do not know what it costs DoD to provide the care. The RAND 2004 child care demand survey indicates that many families of preschool-aged children would prefer to use a CDC. But it is unclear whether they in fact value CDC care at the full cost of providing that care. Parents using FCC may get a subsidy, depending on whether the installation at which they are located offers a subsidy for FCC. Typically, this subsidy is more visible to parents because they know what they would have to pay the provider without the subsidy. The new subsidy programs (Military Child Care in Your Neighborhood, etc.) are also more visible to parents.

The lack of visibility of the CDC subsidy poses a serious barrier to DoD's benefiting from it in terms of recruitment and retention. The services are not unaware of this problem: the U.S. Army, for example, has produced a brochure that clearly states that "the Army covers much of the cost (of CDC care)." Given the size of the subsidy provided, lack of visibility is an issue worthy of serious consideration and possible adjustment. In focus groups conducted as part of our study on child care demand (see Moini, Zellman, and Gates, 2006), many participants expressed the belief that DoD was "making a profit" from CDC care. In other words, parents believe that they pay more for child care than it costs DoD to provide it. On a related point, several CDC users expressed the opinion that if they left the military, it would be easy for them to obtain similar quality child care for the same (or even a lower) price. In fact, it is quite difficult to obtain high-quality child care in civilian centers, and when such care is available, it is much more expensive than the fees paid by military parents for CDC care.

## Summary

The current military child care system focuses most resources on providing in-kind child care benefits through DoD-operated CDCs. The evidence presented in this chapter suggests that a strong emphasis on this one child care option is not effectively meeting the needs of DoD or military parents.

The total cost of CDC care is substantially higher than the cost of FCC or child care provided to DoD by many private contractors. The differences are most significant when looking at care provided to children under two years of age. Of particular note is the cost difference between CDC care and FCC for infants and toddlers, which by our estimate exceeds $7,000 per child per year.

Military families that are fortunate enough to have access to a CDC receive a subsidy that is potentially worth over $10,000 per year per child (depending on the family's income category and the child's age), while families using other child care options receive little or no subsidy. Since many CDC users are completely unaware of the subsidy DoD is providing, DoD is missing an important opportunity to further its goals of recruitment, readiness, and retention through the provision of this valuable subsidy.

Despite the resources being devoted to the CDCs, the CDCs are neither able to serve all military families that need care nor are they able to meet all the child care needs of the military families they serve. Many military families reported in RAND's 2004 survey that child care issues have kept them from reporting to duty or have caused them to miss or be late to work in the past month. Many of these families are CDC users. In addition, even when controlled for family type, CDC users were more likely than other military families to report a propensity to leave the military because of child care issues.

The issues raised in this chapter suggest that DoD may be able to serve more families in a more complete way by shifting resources from CDCs to FCC homes. However, it should not be assumed that this would be an easy shift for DoD to make. Dramatic changes would be needed to make FCC a more attractive option for military families with children. Many parents view CDCs as a safer child care source, as noted in the following chapter, and also as a more stable source of care, since care is not dependent on a single individual, as it is in FCC. Most parents view the idea of a single provider as risky—if she gets sick, parents cannot go to work. While FCC providers are required to file a plan for backup support, no one actually checks on whether the plans are up to date or feasible. The named backup provider is usually another FCC provider (who is eligible to be one because she already has the required background clearance). However, because this provider has her own set of children to care for, oftentimes she will not be able to care for another provider's children if their regular provider is unavailable. If she did, she would exceed the number of children allowed for FCC. Therefore, parents do not trust those plans. This situation is of real concern and has important implications for readiness.

# CHAPTER FIVE

# Transforming Military Child Care

Private-sector efforts to provide child care to employees are typically viewed as direct or indirect tools to improve recruiting, retention, or worker productivity. Many private-sector companies have concluded that reliable child care can contribute to business objectives by improving employee availability, productivity, and retention.

Since the passage of MCCA, most policy discussions related to DoD child care have focused either on the number of families served or the quality of the care provided in DoD programs. Although readiness and retention are often mentioned as a justification for the system, little attention has been paid to the issue of whether and how the system contributes to these goals. Although the child care system theoretically could be used as a recruiting tool, we are aware of no efforts to do so.

There is growing recognition that the system needs to change in order to better address the needs of DoD and of military families. Indeed, the first quadrennial quality of life review (DoD, 2004a) identified child care as a critical quality-of-life issue for military families. But that report noted that the current system is flawed because it is too base centric; the report describes plans to increase capacity through civilian partnerships and subsidies for in-home care.

## The Logic for a Military Child Care Benefit

In this paper, we provide some evidence that child care is a readiness and retention issue. Military members report that child care issues do prevent them from reporting to duty and cause them to be late or absent from work. In addition, some military members indicate that child care issues may lead them to leave the military.

Unfortunately, we know very little about the contribution of the military child care system to recruitment, readiness, and retention goals. In contrast to the many in-depth analyses of specific elements of the military compensation system that have been conducted to determine their effectiveness, there have been no studies that systematically assess the circumstances under which child care issues pose problems for readiness and retention or the extent to which DoD's child care benefits contribute to recruitment, readiness, or retention goals. Given the substantial funds that DoD devotes to child care, an understanding of these relationships is important.

Currently, a huge share of total DoD expenditures on child care is used to provide care in DoD-run CDCs. Even if CDC care were enhancing readiness and retention among those families that use it, the overall effect of the CDCs on these objectives would be limited because so few military families actually use the CDCs. However, there are reasons to doubt that the CDCs are, in fact, having a positive effect on retention even for those families that use them. Parents who use the CDCs were actually more likely than all other families (controlling for family type) to report in our study that they were likely or very likely to leave the military as a result of child care issues (Moini, Zellman, and Gates, 2006). Moreover, the very structure of the CDCs—the fact that they are open a fixed number of hours a day while duty hours may well exceed those hours—limits their effectiveness as a readiness tool.

Even if it were demonstrated that CDCs contribute to employer goals, it is far from clear that provision of this in-kind benefit is meeting employer or family needs in the most cost-effective way. In-kind benefits can be a good policy option when employers can provide a benefit at a lower cost than its value to employees. However, there is no evidence that military child care meets this criterion. Data from our cost survey (Zellman and Gates, 2002) suggest that DoD is no more efficient than other child care providers. In addition, there is no evidence that CDCs are systematically providing care that meets the special needs of military families (such as extended hours or night or weekend care); those needs are often difficult to fill in local communities.

The data are clear that child care needs and preferences vary a great deal across families, which suggests that providing a range of options is more likely to meet these varied needs and to contribute most effectively to readiness and retention. In-kind programs are costly to provide, and, therefore, an employer is unlikely to provide more than one or two options. In-kind programs are best offered in carefully defined situations, such as for late night or overnight care, when care simply might not be available in the community. But DoD is more likely to be successful at meeting a range of family needs if families are able to select the care, from a range of existing options, that best suits their situations.

As a tool to promote retention, in-kind benefits pose another problem, as well: The value of the benefit may not be clear to consumers. Indeed, CDC users, who receive the greatest benefit from the DoD child care system, are often not aware of the value of the benefit they are receiving. It makes little sense for an employer to provide a benefit that employees undervalue. Transparency about the costs of providing CDC care and the value of the subsidy provided to military families through various options should be a basic principle of any new system. Parents may well greet the figures with skepticism: Military parents who have never had to purchase child care in the open market are often shocked to learn how much good-quality care costs there.

The skewed distribution of child care funds, the lack of retention value that DoD is reaping among CDC users, and the opacity of the subsidy value have been of little concern to the child care advocates who designed and built the current system. They know that they have an excellent product and a fairly captive consumer base and thus have not felt the need to "advertise" the quality and value of the benefit. Nor have they made substantial efforts to combat suspicion that DoD is making a profit from CDC fees that many personnel, including CDC users, consider high. It is unfortunate that many military parents explore only DoD-provided child care programs in seeking care, which denies them the opportunity to experience the "sticker shock" that would likely accompany their discovery of the high cost of high-quality civilian-provided child care.

## Reorienting the Child Care System to Better Address DoD Goals

Any transformation of the current military child care system will need to balance the competing priorities of cost, distribution of benefits, and quality. Each of these priorities affects DoD's ability to use the child care system to efficiently leverage its resources to promote readiness, retention, and recruitment. The 2004 first quadrennial quality of life review addresses access (which is one consideration in the distribution of resources) and the need to expand access, but it is silent on the other issues and on the trade-offs that would be involved in expanding access and how access could be expanded in ways that yield an acceptable level of return on DoD's costs. In this section, we discuss each priority and highlight the critical questions for DoD to consider in evaluating and weighting these priorities.

### Cost

As we have discussed in this paper, child care is costly to provide, and DoD CDC care is the most costly child care program for DoD. How much is DoD prepared to spend on child care for military personnel? Is it willing and able to increase spending far beyond current levels, or will any transformation of the military child care system need to hold costs constant?

### Distribution of Benefits

The question of whether and how to provide a child care benefit for military personnel raises important equity and distributional issues. According to the 2003 Status of Forces Survey (DoD, 2004b), 42 percent of military members have children under age 12. DoD already varies service members' compensation according to the number of dependents they have. Should DoD provide an additional benefit that is targeted to child care for these members? Should *all* military members with children under 6 or 12 years of age have access to such a child care benefit, or should a child care benefit be restricted on the basis of family type, deployment status, spouse employment status, or other factors? If DoD believes that a targeted child care benefit is worthwhile, how much should be spent (on a per-military-member basis) to provide this benefit?

Should DoD provide the same level of child care benefits to each military member regardless of the number of children and/or family type? Should single-parent and dual-military families have easier access to child care benefits on the grounds that their readiness is more affected by child care? Should deployed personnel receive a larger benefit on the grounds that their families have less backup care support than other military families? Should the value of the child care benefit vary by child age since we know that it is more costly to care for younger children? Finally, DoD will need to consider whether all families with children of a certain age should be eligible for a child care benefit regardless of the type of care they use (perhaps including parental care). These are all questions that DoD has not considered but that are very relevant to the transformation of the military child care system.

Currently, child care availability is limited by two important factors: DoD's capacity to provide military options and the willingness of military families to use military options. Under the current system, those families that use CDCs receive a substantial subsidy, those that use FCC may receive a small one, but those that either cannot or choose not to use DoD options receive no subsidy at all. Among those least likely to use the military child care system are families living far from a military installation, activated reservists, and military members who are married to civilians. It is also worth noting that because single-parent and dual-military families receive preference for and, therefore, are more likely to use the highly subsidized CDCs, the current system disproportionately benefits these families. Does it makes sense, from a readiness and retention perspective, for DoD to provide these families with a larger per-child subsidy? Our data suggest, for example, that in some single-parent families, the other parent is available to, and does, provide child care. This finding suggests that family type may be too crude an indicator of special child care need. Within the current system, this issue is rarely debated.

## Quality

The current system provides exceptionally high-quality care (which is corroborated by the fact that virtually all CDCs are NAEYC accredited) to a small fraction of military parents who are able to access that care. While this approach most certainly guards DoD against scandal and garners DoD much favorable attention in the child care community, it effectively disregards the quality of care provided to the vast majority of children whose military parents use care in local communities. In addition, the focus on quality of care is divorced from consideration of readiness and retention outcomes. CDCs actually fall short in terms of their ability to provide the flexibility and unusual hours of care that many military families need. It is possible that high-quality child care provides longer-term retention, readiness, and recruiting benefits: If parents recognize and appreciate quality, high-quality care may serve to keep parents in the military and allow them to focus on their work, free of concerns about their children's well-being. If high-quality care improves child outcomes, parents may appreciate this benefit, although the literature on that point is not compelling (Zellman et al., n.d.). Similarly, families that use low-quality child care may face numerous problems that affect readiness and retention, such as low reliability of care, child behavioral problems, and frequent illness. We lack the data to understand these relationships, but we believe that such assessments are worthwhile as a basis for more-informed DoD decisionmaking.

## Presenting and Evaluating Options for Change

The DoD child care system could change in a number of ways. Below, we discuss a range of approaches that DoD might consider to reinvent the DoD child care system and better meet the needs of parents, children, *and* DoD. Which option or set of options should be chosen should be part of a system-level assessment of the key goals that DoD wishes to pursue in providing a child care benefit.

In this paper, we have assumed that at the very least, employer goals of readiness, recruitment, and retention are worth pursuing as key system goals and that a targeted child care benefit is one option for promoting these goals. The discussion of options below is predicated on these assumptions and is focused on assessing the options' potential contributions to readiness, recruitment, and retention. The bottom line in considering these options then, is this question: Can readiness, retention, and recruitment be improved by spending child care dollars differently? If DoD selects a different goal or additional goals, these and perhaps other options must be considered in light of their potential impact on the selected goal or goals. But in any case, we urge DoD to reassess child care and the current system.

**Redistribute Resources Within the Current System, Holding Costs Constant**

In pursuing this strategy, DoD would potentially be able to provide military benefits to more families and/or provide types of care that would have a greater advantage for readiness. A redistribution of resources could involve redirecting money from CDCs to FCC, targeting the child care benefit to different types of families, or focusing the benefit on different types of care.

Shifting resources from CDC care to other types of care, particularly FCC, could improve the system's ability to meet readiness and retention needs. The CDCs that form the backbone of the current system provide care for long hours during the normal workweek. But because they are centers, with high fixed costs associated with each hour they are open, they cannot offer flexibility to a set of users whose work hours are often irregular and may stretch well beyond the normal workweek into nights, weekends, and holidays. FCC, with low fixed costs, can and often does offer such flexibility. Extended hours could become even more prevalent if subsidies were available to providers who offer such care. In addition, because the operating costs per child are much lower in FCC, shifting resources into FCC might allow DoD to serve more children in the system. But as discussed above, the heavy subsidies for CDC care combined with a general preference for CDCs, because of their higher perceived safety (multiple caregivers ensure many opportunities for oversight over the course of the day) and reliability (care is still available if a caregiver is unable to work), propel parents into CDCs.

Dramatic changes would be needed to make FCC a more reliable and flexible option that would support readiness and draw in parents. Several policy options could be considered for improving FCC. DoD could support the development of a network of substitute providers who could stand in for a provider in the event of illness or emergency. Another approach might be to designate a small number of spaces (perhaps enough to cover the children in the care of one FCC provider) in the CDC for FCC children whose provider is not available. Although such changes would increase the cost of FCC, they are unlikely to raise the cost of FCC anywhere near the cost of CDC care. Moreover, they would make FCC a more attractive alternative to those—single parents and dual-military parents—who arguably could benefit most from the greater theoretical flexibility of FCC.

Some provision for care for mildly ill children is another way that the system might better address workforce needs. Although many parents prefer to stay home with sick children, strict CDC policies to protect other children (e.g., no fever over 99 degrees; 24 hour waiting period after an illness) force children whose parents believe they are well enough to go to child care to stay home. This, of course, forces a parent to miss work. It might be useful for DoD to consider whether designating a CDC classroom or subsidizing an FCC home for mildly ill children might be worthwhile.

CDC priorities—given to single-parent and dual-military families—were adopted to ensure these families, which presumably cannot work without child care, access to that care. But this priority may be misplaced: CDC care is also the most rigid type of care, which may not be the best match for families that may not be able to rely on a second parent to provide backup care. Our data also suggest that this care does not support retention decisions.

Rethinking priority policies from the perspective of *both* child care need and the degree to which care characteristics fit with likely DoD and service member needs would be an important way to revise the system. For example, it might be far more effective in terms of readiness to give single-parent and dual-military families FCC priority and also provide higher subsidies to FCC providers who care for children from these family types and who agree to provide extra hours of care as needed (within limits). This change would lower the relative cost of FCC for these families, making it more attractive. The assurance of extended care hours would make it much easier for them to fulfill their work requirements, be less disruptive to the work group, and increase their children's continuity of care, which is often a problem for children of single parents and dual-military members cared for in CDCs.

DoD may also wish to redistribute resources in order to allocate child care benefits more equitably. To do this without increasing overall expenditures would involve a reallocation from those who are currently receiving a large subsidy to those who are receiving little or no subsidy. Although it is always difficult for an employer to remove or reduce existing benefits, it might be relatively easy to accomplish in the child care arena because as children grow up, the "winners" (those with young children) quickly become "losers" (those with older children) since CDC infant care is the most heavily subsidized.

Alternatively, DoD may wish to redistribute child care benefits to target them either to those families that value them the most, such as military members who are deployed, or to the families of those DoD values most highly, such as individuals with special skills. There is ample precedent in the private sector for using child care benefits in this way (see Chapter Two).

However, shifting resources from CDCs to other care options may prove to be a challenge, in part due to the requirements of MCCA. Zellman and Gates (2002) found that, in 1998, with the exception of the U.S. Navy, each service was operating very close to budget constraints imposed by MCCA: About half of CDC expenditures were nonappropriated fund expenditures (i.e., money that came from parent fees) and half were appropriated fund expenditures (i.e., a subsidy from DoD). Only the Navy, in which nonappropriated fund expenditures covered only 34 percent of total expenditures, seemed to be providing a larger subsidy than required by law. Unless this funding pattern has changed dramatically in the past ten years, the two main ways that DoD could reduce CDC expenditures without a change in MCCA would be either to reduce the total number of CDC slots or to increase the share of preschool-aged children served in the CDCs, shifting infants to FCC.

Although closing a CDC is not likely to be a viable option in general, over the next several years, 22 major DoD installations are scheduled to be closed through execution of the recommendations of the 2005 Base Realignment and Closure (BRAC) process. As a result, at least ten CDCs, serving approximately 1,200 children, will be closed as well. Installations anticipating an increase in population as a result of BRAC have proposed plans for the construction of new CDCs, including the addition of more slots. For example, Fort Riley, Kansas, has a $5.7 million CDC project planned that would provide care for 198 children. However, the *Fiscal Year 2007 Continuing Resolution* funding bill did not include the full funding for the proposed construction projects related to closures and realignments. As a result, it is uncertain what the ultimate net gain or loss in CDC slots will be ("Governor's Military Council Votes," 2007). DoD could use the execution of BRAC recommendations as an opportunity to reallocate resources to alternative child care approaches, such as an expansion of the Military Child Care in Your Neighborhood Program, an expansion of FCC subsidies, or additional spending on R&R services (including negotiated discounts with major providers).

### Expanding the Military Child Care Benefit

In support of recruitment, readiness, and retention goals, DoD may wish to expand the child care benefit to cover more military families and a broader set of child care needs. For example, military members with a civilian spouse who does not work outside the home are not currently eligible for any military child care benefits, nor do they have access to DoD CDCs, except on an hourly space-available basis. DoD may wish to offer such families access to CDCs or other forms of child care subsidies, particularly when the military spouse is deployed. DoD could expand the child care benefit in a number of ways. To begin, it could subsidize some of the supplemental types of child care discussed above (mildly ill child care or care to cover night and weekend duty) while continuing to provide the care as it currently does. Another option would be to expand the current system of DoD-provided care so that it is available to more military families. Lastly, DoD could expand the types of care that are eligible for subsidies. Any such expansion would, of course, involve additional resource expenditure. For this reason, DoD would want to forecast as carefully as possible the benefit it likely would receive from providing a targeted military child care benefit as it considers one or more of these options.

## Expanding the Current System of DoD-Provided Care

DoD could consider expanding DoD-provided care and evaluate the system in terms of availability of care and contribution to readiness. This undertaking would likely be costly, not only in terms of operating costs, but also military construction costs. Cost could be moderated by expanding use of FCC homes and moving the care of young children out of CDCs, focusing center-based care on older children. However, even if enough FCC providers stepped forward, it is unlikely that expanded CDC care or FCC would meet the needs of all military families, many of whom live far from the installation. Therefore, the ability of this approach to meet the needs of military families is inherently limited. This model assumes the quality of care provided to children would continue to be high. With services widely available, DoD could begin to track the contribution of child care services to readiness and retention through surveys. At a minimum, DoD would need to make military parents aware of the value of the subsidy they are receiving. It would also need to consider many of the issues raised in the previous section and might still want to alter the ways in which FCC and CDC care are used, directing a larger share of resources to FCC expansion.

According to the first quadrennial quality of life review (DoD, 2004a), DoD is currently delivering child care to approximately 175,000 military children through its varied child care settings. The report articulates a goal of increasing the number of spaces to 215,000. Given that over 500,000 military members have dependents under age 12 and many have more than one child in that age range, it is clear that even the expanded system will not come close to serving all military children under 12 years of age.

## Expand Access to Child Care Through Vouchers

DoD could expand access to child care through the use of cash benefits, vouchers, and/or negotiated discounts with local providers, while continuing to provide some amount of FCC and CDC care. There would be many challenges and questions to be resolved in pursuing such an option. This option would likely be costly. DoD would need to determine which military members would be eligible for the voucher benefit and how to value the benefit, i.e., determine how much the voucher would be worth. DoD would also need to determine what quality criteria a provider would need to meet in order to qualify to receive military vouchers. In most communities in the country, the availability of high-quality child care is limited, as discussed above. DoD must recognize that a voucher system would stretch local capacity and that it would take time for new providers—particularly high-quality providers—to emerge. However, this option would expand access to child care benefits and would increase equity. In the long term, it would also likely increase child care availability and the average quality of care that

DoD dependents receive. The quality of care provided could be regulated through the setting of quality criteria for subsidy eligibility, which would ensure an acceptable level of quality for more families as well. If DoD believes that there is an association between care quality and readiness, increasing the aggregate quality level of care is important.

In principle, vouchers or cash payments could be used for either DoD-provided or off-base care; if DoD were to offer vouchers, there are strong arguments for folding the current DoD-provided options into any voucher system. Under an integrated voucher approach, DoD would continue to provide care in CDCs and FCC, but parental subsidies would not be tied to the type of care used. Instead, parents would get a voucher that they could use to purchase child care in CDCs, FCC, or approved off-base arrangements. CDCs and FCC, like private providers, would accept the voucher as partial (or perhaps full) payment for child care. A major value of this approach is that it would improve the visibility of the child care subsidy that parents receive from DoD. These vouchers could ultimately be used in the context of a cafeteria benefit plan. Depending on how CDC care was costed under a revamped system, this approach could also provide DoD with better information about the value that parents place on different types of care. A major problem with in-kind benefits is that people will use them even if the value of the benefit is less than the cost to the employer of providing the benefit. By subsidizing only one (or sometimes two) types of child care, DoD inflates demand for that type of care.

Within the context of an integrated voucher system, parents would continue to use CDCs to the extent that parents understand and value the quality of care that CDCs provide and appreciate the convenience of on-base care. Under such a system, any waiting list for CDC care would provide a more reliable signal of excess demand than is true under the current system, in which the disproportionate CDC subsidy inflates demand for that care.

In implementing a voucher system, DoD would need to grapple with the questions of how to value the voucher and how to price CDC care. Each of these issues is discussed briefly below.

**How to Value the Voucher?** A key policy decision that would confront DoD under a voucher system is how to value the voucher and whether to vary that value by family characteristics, geographic location, or cost of care. The value of the voucher could vary by child age and family income (or not). It could vary with the cost of living in the local area (as with cost of living allowances). It could be based on the market rate for child care in the local area, following the model of the Basic Allowance for Housing, which varies according to the local market rate for housing. DoD might choose to give the military member one voucher per child or provide one voucher per military member. Dual-military families might be eligible for two vouchers per child (one for each military member). As mentioned earlier, DoD would need to consider whether all military families are eligible or only those who are single parents or have an employed spouse. These are policy choices that DoD would need to think through in determining how to allocate a child care entitlement.

**How to Price CDC Care?** Under a comprehensive voucher system, FCC providers would still be free to set their own fees and could accept the voucher as partial or full payment for child care. Pricing for CDC care would need to be completely revamped. Private-sector centers typically tie fees to child age, which reflects the very different costs associated with providing care under different staff-to-child ratios. Typically, care for infants costs up to 50 percent more than care for preschool-aged children; care for toddlers may cost up to 30 percent more than care for preschoolers. DoD would need to consider varying fees by child age, something it has resisted in the past because it is the youngest, lowest-ranked members who have the youngest children. A fee structure in which parents' fees are higher for younger children would provide parents with an incentive to select more cost-effective care. One likely effect would be to encourage more families with infants and pretoddlers to use FCC, which tends to be less costly to provide than CDC care for the youngest children. Those parents for whom the CDC structure is highly attractive and convenient may be willing to assume the higher parent fees and send even their youngest children to the centers; others might choose an FCC home or other type of provider instead.

**Other Changes/Activities.** DoD may want to provide additional support for child care that directly affects readiness (such as child care for weekend drills or overnight duty). A voucher system could easily accommodate such a support policy. Military members simply could be granted a voucher supplement in the event of such demands.

**Capacity of FCC and Off-Base Providers to Care for Infants and Pretoddlers.** As stated above, infant and pretoddler care is the most costly to provide. Because FCC providers who care for infants and pretoddlers are limited in the number of other children they can care for, FCC providers should be allowed, and even encouraged, to charge higher rates for infants and toddlers.[1] This price difference could be offset by a higher voucher value for the youngest children. DoD may need to invest some resources in order to increase the capacity of FCC and off-base providers.

## Improvement of Local Provider Quality

Although we identified no studies that have examined the effect of child care quality on readiness, retention, or child outcomes in the military context, it is certainly plausible that high-quality care can contribute to the well-being of military children (and by extension the military family) and to readiness and retention; this belief is widely viewed as a reason to provide high-quality care in CDCs and FCC. Parents who have access to reliable, high-quality care may be more able to focus on their work and also may be more likely to have well-adjusted children who are less prone to behavioral problems or to illness and, therefore, the parents may be less likely to miss work or leave the military. This argument suggests that DoD might benefit if military families use high-quality, community-based child care options.

If DoD chooses to focus on improving local provider quality, its considerable numbers and purchasing power could exert a positive influence on the overall quality of care in communities surrounding military installations. This influence could be accomplished in a number of ways, as discussed below.

One straightforward approach would be to offer subsidies in which the subsidy amount is tied to a particular quality standard. This subsidy would reward higher levels of quality and incentivize community providers to achieve higher quality. If the provider has excess slots (which does not occur often in providers of reasonable quality) and/or sees the DoD subsidy as an incentive to achieve a quality goal, providers would be motivated to achieve the DoD standard. Selecting the standard must be done carefully: A tough standard such as NAEYC accreditation is likely to incur substantial, ongoing costs, particularly if staff-to-child ratios need to be reduced. Any subsidy must be sufficient to cover costs incurred to meet standards or it will not have the desired incentive value for providers.

---

[1]  Such a change would be much easier to implement in conjunction with a change to the CDC fee schedule as described above.

It may be necessary for DoD to do more than simply offer a subsidy to ensure adequate quality. It may be necessary to provide some support for quality improvement. These quality-improvement services might take the form of quality ratings and feedback alone, training tied to the feedback, staff scholarships, or support for facility improvements. Obviously, before DoD would enter into this level of involvement with civilian providers, there would need to be a contract specifying provision of services to military dependents.

Another approach to improving quality might involve directly working with parents. For example, DoD might need to be far more active in promoting parent education about child care quality. Selecting and overseeing a QRS and developing a broader subsidy program would also support military personnel using civilian care. That program would need to be carefully designed and managed so that the subsidies—through the establishment of standards, a rating system, tiered reimbursement, or other means—meet children's, families', and DoD's needs.

**Move Away from Regulation**

Over the years, DoD has run a highly regulated system in order to comply with MCCA mandates, as discussed above. Many of the changes described in this section would require DoD to adopt a very different approach to managing the child care benefit: one that creates a system with checks and balances, that monitors the contribution of system elements to DoD and family goals, and that does not depend so heavily on regulation. Such a system would provide incentives for quality improvement by providers; it would provide parents with vouchers or subsidies and easily understood quality information; then it would allow parents to make choices that affect provider outcomes.

**Dependent Care Reimbursement Accounts**

Independent of any changes to the overall child care system, DoD could consider offering DCRAs to military members, as described in Chapter Two. However, despite the popularity of these accounts among private-sector employers, we question whether many military members would find them to be a useful benefit, given current restrictions on their use, and caution that it might raise risks for DoD. For military families that face a tremendous amount of uncertainty about the course that their lives will take over a year's time, it may not make sense to commit funds to this type of account, particularly when members stand to lose unspent funds remaining at the end of the year, which is a feature of these accounts. For example, if a military member is deployed and the spouse decides to quit her job, the family could lose the money deposited into the account. DoD could reduce this risk by securing changes to the tax code that defines the set of qualifying "life events" that allow families to make changes to their DCRA elections. Specifically, the list could be expanded to include a permanent change of station and deployment. Finally, for lower-income families, the child care tax credit provides the same tax benefits as the DCRA without the risks. The DCRA would primarily benefit mid-career and senior military members. In the end, DCRAs (with some revisions to account for military-specific considerations) could be a useful and low-cost benefit for DoD to provide to mid-career and senior military members, but DoD should proceed carefully with this benefit to avoid a possible outcry from military families that might lose money that they had deposited in the account based on assumptions about their plans that were altered by military needs.

## Recommendations for Assessing the Options

The previous section provides many options for transforming the military child care system. To choose among these options, DoD needs to consider its goals and assess the evidence regarding the military child care system's contribution to those goals. Unlike other elements of DoD compensation, which have been and continue to be analyzed in detail, we know very little about the contribution of military child care to readiness, retention, recruitment, and even child outcomes. Aside from periodic customer satisfaction-type questions included on DoD's active duty survey, DoD collects and retains little information about the child care system and its contribution to DoD goals. DoD should centrally record and track information about who is using the child care system and the amount and type of care being used. This information could then be used to assess the impact of system utilization on readiness and retention. Ideally, DoD would also retain information on the children being served so that assessments of near-term child outcomes (e.g., health status, test scores from first grade) could be conducted. Assessments of long-term outcomes could also be considered, including the extent to which these children eventually choose to join the military themselves.

DoD might also want to consider conducting a regular child care needs assessment in the face of ever-changing circumstances. For example, new studies reveal that child care needs may increase for families with a deployed member. If deployment levels are increasing, DoD may need to expand its support for child care in new ways, e.g., respite care.

As described in this paper, there are several options that DoD could consider for revamping the military child care benefit either to redirect resources to different types of families and/or different child care options or to expand access to child care benefits. To be able to fully consider options for system change, DoD would need to conduct targeted studies of the child care market in local military communities. These data would help DoD to understand what types of care military families are using, how much they are paying, the range of quality currently available, the extent of military parents' knowledge about quality, and what steps DoD may need to take to educate parents about this important aspect of child care. It would also provide input to a key DoD decision that the department might need to make about its degree of involvement in promoting the development and use of high-quality care.

Once collected, DoD should make these data available in a user-friendly form to recruiters. This step would be part of a larger effort that DoD should support to communicate the value of the DoD child care system to military members. In addition to communicating this information directly to military members, commanding officers might use it to encourage members to reenlist.

DoD has been distributing a valuable benefit to service members with little concern about equity, and it has forgone rich opportunities to educate parents about the quality of the military child care system or to use its influence to improve child care in local communities. DoD can do a better job of addressing the child care needs of its military families and its own needs for a stable, ready force in the face of ever-changing circumstances. Many of the ideas presented here are provocative; many may be greeted initially with suspicion. Most of these ideas will require DoD to change what the military child care system does. But we believe that considering these options will be a worthwhile exercise for DoD and the families it serves.

# References

BLS—*see* U.S. Department of Labor.

Bond, James T., Ellen Galinsky, Stacy S. Kim, and Erin Brownfield, *2005 National Study of Employers,* New York, N.Y.: Families and Work Institute, 2006. As of February 2008:
http://www.familiesandwork.org/eproducts/2005nse.pdf

Bond, James T., Ellen Galinsky, and J. E. Swanbert, *The 1997 National Study of the Changing Workforce,* "Executive Summary," New York, N.Y.: Families and Work Institute, 1998.

Bond, James T., Cindy Thompson, Ellen Galinsky, and David Prottas, *Highlights of the National Study of the Changing Workforce,* No. 3, New York, N.Y.: Families and Work Institute, 2002.

Bright Horizons, *The Real Savings from Employer Sponsored Child Care: Investment Impact Study Results,* Watertown, Mass.: Bright Horizons Family Solutions, 2004. As of June 2007:
http://www.brighthorizons.com/site/pages/InvestmentImpact.pdf

Buddin, Richard, *Building a Personnel Support Agenda: Goals, Analysis Framework, and Data Requirements,* Santa Monica, Calif.: RAND Corporation, MR-916-OSD, 1998. As of February 2008:
http://www.rand.org/pubs/monograph_reports/MR916/

Buddin, Richard, Carole Roan Gresenz, Susan D. Hosek, Marc N. Elliott, Jennifer Hawes-Dawson, *An Evaluation of Housing Options for Military Families,* Santa Monica, Calif.: RAND Corporation, MR-1020-OSD, 1999. As of February 2008:
http://www.rand.org/pubs/monograph_reports/MR1020/

Congressional Budget Office (CBO), *Evaluating Military Compensation,* Washington, D.C.: CBO, June 2007.

DoD—*see* U.S. Department of Defense.

Federal Employee Education & Assistance Fund (FEEA), Web site, Littleton, Colo.: FEEA, n.d. As of June 22, 2007:
http://www.feea.org/childcare/ccadmin.html

GAO—*see* U.S. Government Accountability Office.

Gates, Susan M., Gail L. Zellman, and Joy S. Moini, unpublished research on demand for child care among military families: insights from focus groups, Santa Monica, Calif.: RAND Corporation, n.d.

———, *Examining Child Care Need Among Military Families,* Santa Monica, Calif.: RAND Corporation, TR-279-OSD, 2006. As of February 2008:
http://www.rand.org/pubs/technical_reports/TR279/

Golfin, Peggy A., *Toward an Understanding of the Role of Incentives in Enlisted Recruiting,* Alexandria, Va.: Center for Naval Analyses, 2003. As of February 2008:
http://www.cna.org/documents/D0007706.A2.pd

"Governor's Military Council Votes to Support Pat Roberts' Effort to Restore Kansas Military Funding," States News Service, February 8, 2007.

Greenberg, M., "Next Steps for Federal Child Care Policy," *The Future of Children,* Vol. 17, No. 2, Fall 2007, pp. 73–96.

Hansen, Michael L., Jennie W. Wenger, and Anita Hattiangadi, *Return on Investment of Quality of Life Programs,* Alexandria, Va.: CNA Corporation, CRM D0006807.A2, October 2002.

Hattiangadi, Anita U., *Private Sector Benefit Offerings in the Competition for High-Skill Recruits,* Alexandria, Va.: CNA Corporation, CRM D0003563.A2, December 2001.

Hogan, Paul F., "Overview of the Current Personnel and Compensation System," in Cindy Williams, ed., *Filling the Ranks: Transforming the U.S. Military Personnel System,* Cambridge, Mass.: MIT Press, 2004, Chapter 2.

Moini, Joy S., Gail L. Zellman, and Susan M. Gates, *Providing Child Care to Military Families: The Role of the Demand Formula in Defining Need and Informing Policy,* Santa Monica, Calif.: RAND Corporation, MG-387-OSD, 2006. As of February 2008:
http://www.rand.org/pubs/monographs/MG387/

Murray, Carla T., "Transforming In-Kind Compensation," in Cindy Williams, ed., *Filling the Ranks: Transforming the U.S. Military Personnel System,* Cambridge, Mass.: MIT Press, 2004, Chapter 9.

National Association of Child Care Resource & Referral Agencies (NACCRRA), "Military Partnerships," Arlington, Va.: NACCRRA, n.d. Accessed November 2, 2007:
http://www.naccrra.org/MilitaryPrograms/

———, *Breaking the Piggy Bank: Parents and the High Price of Child Care,* Arlington, Va.: NACCRRA, 2006.

———, *Air Force/NACCRRA Quality Family Child Care (QFCC) Project: Final Report,* Washington, D.C.: NACCRRA, January 2007.

OPM—*see* U.S. Office of Personnel Management.

Overturf Johnson, Julia, *Who's Minding the Kids? Child Care Arrangements: Winter 2002,* Current Population Reports, P70-101, Washington, D.C.: U.S. Census Bureau, 2005.

Peisner-Feinberg, E. S., M. R. Burchinal, R. M. Clifford, N. Yazejian, M. L. Culkin, J. Zelazo, C. Howes, P. Byler, S. L. Kagan, and J. Rustici, *The Children of the Cost, Quality, and Outcomes Study Go to School: Technical Report,* Chapel Hill, N.C.: University of North Carolina, Frank Porter Graham Child Development Center, 1999.

Press, Eyal, "Family-Leave Values," *New York Times Magazine,* July 29, 2007, pp. 36–41.

Raezer, Joyce W., "Transforming Support to Military Families and Communities," in Cindy Williams, ed., *Filling the Ranks: Transforming the U.S. Military Personnel System,* Cambridge, Mass.: MIT Press, 2004, Chapter 10.

Rostker, Bernard D., *I Want You! The Evolution of the All-Volunteer Force,* Santa Monica, Calif.: RAND Corporation, MG-265-RC, 2006. As of February 2008:
http://www.rand.org/pubs/monographs/MG265/

Segal/Sibson, *Rewards of Work Study: How Employees Gauge the Rewards of Their Work,* Cleveland, Ohio: The Segal Group, Inc., 2006a. As of February 2008:
http://www.segalsibson.com/publications/surveysandstudies/2006ROWno1.pdf

———, *Rewards of Work Study: Insights into Employee Engagement,* Cleveland, Ohio: The Segal Group, Inc., 2006b. As of February 2008:
http://www.segalsibson.com/publications/surveysandstudies/2006ROWno2.pdf

———, *Rewards of Work Study: Benefit Satisfaction, Retention and Productivity,* Cleveland, Ohio: The Segal Group, Inc., 2006c. Abstract, as of February 2008 (PDF available via email):
http://www.sibson.com/corporate/pub-corporate.cfm?ID=736

Shellenback, K., *Child Care & Parent Productivity: Making the Business Case,* Ithaca, N.Y.: Cornell University, Cornell Cooperative Extension, 2004.

Shellenbarger, Sue, "New Ideas for Perennial Issues, from Day Care to Job Shares," *Wall Street Journal,* April 12, 2007, p. D1.

Strawn, Thomas M., "The War for Talent in the Private Sector," in Cindy Williams, ed., *Filling the Ranks: Transforming the U.S. Military Personnel System,* Cambridge, Mass.: MIT Press, 2004, Chapter 4.

U.S. Department of Defense, *Report of the Ninth Quadrennial Review of Military Compensation,* Washington D.C.: DoD, March 2002.

———, *Report of the 1st Quadrennial Quality of Life Review,* Washington D.C.: DoD, May 2004a. As of February 2008:
http://www.militaryhomefront.dod.mil/portal/page/itc/MHF/MHF_DETAIL_1?content_id=168185

———, Defense Manpower Data Center, *Background and Characteristics of Military Families: Results from the 1992 DoD Surveys of Officers and Enlisted Personnel and Military Spouses,* Arlington, Va.: Defense Manpower Data Center, Report No. 97-002, April 1997.

———, Defense Manpower Data Center, *November 2003 Status of Forces Survey of Active-Duty Members: Tabulation of Responses,* Arlington, Va.: Defense Manpower Data Center, Report No. 2003-031, March 2004b.

———, Defense Manpower Data Center, *Selected Manpower Statistics, Fiscal Year 2005,* Arlington, Va.: Defense Manpower Data Center, 2005. As of November 13, 2007:
http://siadapp.dmdc.osd.mil/personnel/M01/fy05/m01fy05.pdf

———, *Military Family Resource Center, Career Resource Library,* "Overview of Military Child Development System," Minneapolis, Minn.: CareerOneStop, 2001. As of November 13, 2007:
http://www.acinet.org/crl/CRL_RRSearch.aspx?docn=6509&LVL1=9&LVL2=51&LVL3=n&CATID=600&PostVal=2

U.S. Department of Labor, Bureau of Labor Statistics (BLS), *Issues in Labor Statistics,* "Employer-Sponsored Childcare Benefits," Summary 98-9, August 1998. As of February 2008:
http://www.bls.gov/opub/ils/pdf/opbils24.pdf

———, *National Compensation Survey: Employee Benefits in Private Industry in the United States, 2002–2003,* Bulletin 2573, January 2005a. As of February 2008:
http://www.bls.gov/ncs/ebs/sp/ebbl0020.pdf

———, *National Compensation Survey: Employee Benefits in Private Industry in the United States, March 2004,* Summary 04-04, 2004. As of February 2008:
http://www.bls.gov/ncs/ebs/sp/ebsm0002.pdf

———, *National Compensation Survey: Employee Benefits in Private Industry in the United States, March 2005,* Summary 05-01, August 2005b. As of February 2008:
http://www.bls.gov/ncs/ebs/sp/ebsm0003.pdf

———, *National Compensation Survey: Employee Benefits in Private Industry in the United States, March 2006,* Summary 06-05, August 2006. As of February 2008:
http://www.bls.gov/ncs/ebs/sp/ebsm0004.pdf

———, *National Compensation Survey: Employee Benefits in Private Industry in the United States, March 2007,* Summary 07-05, August 2007. As of February 2008:
http://www.bls.gov/ncs/ebs/sp/ebsm0006.pdf

U.S. General Services Administration (GSA), *Child Care,* "Frequently Asked Questions," New York, N.Y.: Child Care Program Office, GSA, n.d. As of June 19, 2007:
http://www.gsa.gov/Portal/gsa/ep/contentView.do?faq=yes&pageTypeId=10430&contentId=8355&contentType=GSA_OVERVIEW

U.S. Government Accountability Office (GAO), *Military Personnel: Preliminary Results of DoD's 1999 Survey of Active Duty Members,* Washington, D.C.: GAO, GAO/T-NSIAD-00-110, March 8, 2000.

———, *An Assessment of Dependent Care Needs of Federal Workers Using the Office of Personnel Management's Survey,* Washington, D.C.: GAO, GAO-07-437R, March 30, 2007. As of February 2008:
http://www.gao.gov/new.items/d07437r.pdf

U.S. Office of Personnel Management (OPM), "Child Care Subsidy Program," Washington, D.C.: OPM, n.d. As of May 2, 2007:
http://www.opm.gov/Employment_and_Benefits/WorkLife/FamilyCareIssues/ChildCare_Subsidy/index.asp

————, *Guide for Implementing Child Care Legislation,* "Public Law 107-67, Section 630," Washington, D.C.: OPM, October 2004. As of May 2, 2007:
http://www.opm.gov/Employment_and_Benefits/worklife/officialdocuments/handbooksguides/childcare_legislation/index.asp

Unum Provident, *Buyers Study: The Trends Employers Are Facing and the Plans They're Buying,* Portland, Me.: Unum Provident, 2007a. As of February 2008:
http://www.unum.com/buyerstudy/

Unum Provident, *Buyers Study: Supplement: An Insider's View to Industries,* Portland, Me.: Unum Provident, 2007b. As of February 2008:
http://www.unum.com/buyerstudy/

*Wall Street Journal,* "Career Journal" Web page, n.d. As of February 2008:
http://careerpath.org/columnists/workfamily/20020920-workfamily.html

Watson Wyatt Worldwide, *Watson Wyatt's Human Capital Index: Human Capital as a Lead Indicator of Shareholder Value: 2001/2002 Survey Report,* Washington, D.C.: Watson Wyatt Worldwide, n.d.

Williams, Cindy, ed., *Filling the Ranks: Transforming the U.S. Military Personnel System,* Cambridge, Mass.: MIT Press, 2004.

"2006 100 Best Companies," *Working Mother,* n.d. As of July 18, 2007:
http://www.workingmother.com/web?service=vpage/77

Zellman, Gail L., and Susan M. Gates, *Examining the Cost of Military Child Care,* Santa Monica, Calif.: RAND Corporation, MR-1415-OSD, 2002. As of February 2007:
http://www.rand.org/pubs/monograph_reports/MR1415/

Zellman, Gail L., Michal Perlman, Vi-Nhuan Li, and Claude Setodji, unpublished research on improving child care quality through quality rating systems: assessing the validity of the Qualistar Early Learning quality rating improvement system, Santa Monica, Calif.: RAND Corporation, n.d.